FUNDAMENTALS OF COMPUTER-AIDED CIRCUIT SIMULATION

THE KLUWER INTERNATIONAL SERIES
IN ENGINEERING AND COMPUTER SCIENCE

VLSI, COMPUTER ARCHITECTURE AND
DIGITAL SIGNAL PROCESSING

Consulting Editor
Jonathan Allen

Other books in the series:

FUNDAMENTALS OF COMPUTER-AIDED CIRCUIT SIMULATION

William J. McCalla
Hewlett-Packard

Kluwer Academic Publishers
Boston/Dordrecht/Lancaster

Distributors for North America:
Kluwer Academic Publishers
101 Philip Drive
Assinippi Park
Norwell, MA 02061, USA

Distributors for the UK and Ireland:
Kluwer Academic Publishers
MTP Press Limited
Falcon House, Queen Square,
Lancaster LA1 1RN, UNITED KINGDOM

Distributors for all other countries:
Kluwer Academic Publishers Group
Distribution Centre
Post Office Box 322
3300 AH Dordrecht, THE NETHERLANDS

Library of Congress Cataloging-in-Publication Data

McCalla, William J.
  Fundamentals of computer-aided circuit simulation.

  (The Kluwer international series in engineering
and computer science ; #37)
  Includes index.
  1. Integrated circuits—Very large scale integration—
Design and construction—Data processing.   2. Computer-
aided design.   I. Title.   II. Series.
TK7874.M355 1988   621.381 '93   87–22846
ISBN 0–89838–248–3

CONTENTS

LIST OF FIGURES

LIST OF TABLES

PREFACE

From little more than a circuit-theoretical concept in 1965, computer-aided circuit simulation developed into an essential and routinely used design tool in less than ten years. In 1965 it was costly and time consuming to analyze circuits consisting of a half-dozen transistors. By 1975 circuits composed of hundreds of transistors were analyzed routinely. Today, simulation capabilities easily extend to thousands of transistors. Circuit designers use simulation as routinely as they used to use a slide rule and almost as easily as they now use hand-held calculators. However, just as with the slide rule or hand-held calculator, some designers are found to use circuit simulation more effectively than others. They ask better questions, do fewer analyses, and get better answers. In general, they are more effective in using circuit simulation as a design tool. Why?

Certainly, design experience, skill, intuition, and even luck contribute to a designer's effectiveness. At the same time those who design and develop circuit simulation programs would like to believe that their programs are so easy and straightforward to use, so well debugged and so efficient that even their own grandmother could design effectively using their program. Unfortunately such is not the case and blame does not rest on their grandmother! Thus it must be said that any designer, regardless of his or her innate ability, will design more effectively using computer simulation, if he or she understands more fully how the program works, how it manipulates data, and consequently, what are its strengths and its limitations.

This text is intended as both an introduction to and as a quick summary of those numerical techniques which have been found to be relevant to circuit simulation. As such it should be suitable for use by advanced undergraduate students, graduate students, and practicing engineers alike. At least a rudimentary understanding of calculus (*e.g.* derivatives and Taylor series), linear algebra (*e.g.* systems of linear equations), numerical analysis (*e.g.* nonlinear equation solution and numerical integration), and basic circuit and semiconductor device theory are desirable for but not essential to an understanding of the material presented herein.

In the following chapters, the formulation and implementations of most of the various numerical techniques and algorithms which have been found useful in circuit simulation programs are described. While it has not been the intention to prove that any one method or approach is better than others, various methods are compared where possible. Some of these methods have been used extensively, while others have not been used at all. The intent is to cover this ground at an intuitive and hopefully understandable level. While mathematical accuracy has been considered desirable, mathematical rigor in derivations has sometimes been sacrificed in favor of plausibility arguments.

Chapter 1 outlines the basic steps involved in equation formulation from branch constitutive equations through nodal, modified nodal and sparse tableau formulations. Chapter 2 treats the solution of systems of linear algebraic equations by Gaussian elimination, LU factorization and other variations including iterative methods. The evaluation of determinants is also covered. Sparse matrix methods including storage and pivot ordering are covered in Chapter 3. In Chapter 4, equation solution methods are extended to nonlinear systems. Newton-Raphson iteration, its convergence and termination, and its variations are described. In addition, an experimental technique for suppressing the explicit equations normally associated with internal device nodes is presented. Numerical integration techniques are covered in Chapter 5 from the viewpoints of their application, their construction, their error properties and stability properties and finally automatic timestep control. Chapter 6 treats the adjoint network and its use in computing circuit sensitivities. Noise and delay calculations are also considered. Pole-zero evaluation using Muller's method is described in Chapter 7. Finally in Chapter 8, an approach to the statistical description of circuits for simulation is presented.

The material presented is somewhat generic in that no specific simulation program is used throughout as an example. Formulation of circuit equations, a task basic to all simulators, and representation of circuit equations as a sparse system are presented first as each of the later chapters presume the existence of circuit equations. The next two chapters each address the solution of relevant simulation sub-problems yet are prerequisites to addressing the solution of the large-signal time-domain simulation problem as treated in Chapter 5. By way of explanation, at each new timepoint in a simulation, the characteristic non-linear differential equations are converted to an equivalent set of non-linear algebraic equations through numerical integration techniques. The derived set of non-linear algebraic equations are solved through an iterative process which at each iteration results in a system of linear algebraic equations. The last three chapters on sensitivity analysis, pole-zero analysis, and statistical analysis are relatively independent of each other and need not be treated in order.

This material was originally assembled into its present form for a seminar taught by the author at the University of California at Berkeley in 1975 and again the following year both at UC Berkeley and at Stanford University. Though more than ten years have passed since the material was originally assembled, both content and derivation remain relevant to today's simulators. Some recent innovations such as relaxation-based algorithms are not treated here but are described in detail in other recent texts. The techniques presented herein remain fundamental to such topics. It is hoped that the material presented in these eight chapters will, in fact, aid in the development of insight into the entire field of circuit simulation and prove useful to both the design engineer and the program developer.

The author would like to take this opportunity to express his appreciation and thanks to the many people who have contributed along the way to his career: to his parents, Jim and Ida McCalla, for having provided him his start in life, for having fostered a love for learning and for having instilled a spirit of persistence; to Don Pederson, his research advisor at UC Berkeley, for having had the faith to accept him as a graduate student and the patience to wait for worthwhile results; to Ivan Cermak, his manager at Bell Telephone Laboratories, who was willing to hire him and then let him continue to pursue his interest in simulation; to Leo Craft, his manager at Signetics, who allowed him the time to prepare and teach the initial seminar; to Merrill Brooksby and the other managers at Hewlett-Packard Company for providing additional encouragement and an invigorating environment in which to grow; and certainly to his wife Judi and sons Matthew and Justin for their willingness to put up with his odyssey and for their continued support over many years. There are many other colleagues, peers, former students, etc. from both industry and academia who should also be acknowledged individually if only time, space, and memory permitted. Most will know who they are. So to both those who do know as well as to those who don't, the author also expresses his appreciation and thanks .

1. CIRCUIT EQUATION FORMULATION

1.1 Branch Constitutive Equations

Virtually all presently available circuit analysis programs start from the same point, an elemental circuit description. This description of circuit elements and their interconnections is converted by the programs into a set of circuit equations. Before considering a set of circuit equations collectively, it is useful to consider the branch constitutive equations of individual circuit elements. Further, it is useful to distinguish the forms of the defining branch relations for quiescent dc, time-domain or transient and small-signal ac analyses. The branch constitutive equations for a number of basic elements are given in Table 1.1. As they constitute the basis for 90% of all computer-aided circuit simulation, they are all that will be considered at this point. It will later be shown that nonlinear elements such as diodes and transistors can virtually always be modeled in terms of these basic relationships.

1.2 Nodal Analysis

Because of its fundamental simplicity nodal analysis is one of the oldest and most widely used methods of formulating circuit equations for computer simulation. It can easily be shown that the validity of the nodal equation formulation, as with all other formulations, is based on the application of:

1) Kirchoff's Current Law (KCL).

2) Kirchoff's Voltage Law (KVL).

3) Branch Constitutive Equations (BCE).

The particular emphasis of nodal analysis is the use of Kirchoff's Current Law (KCL) in writing an equation for the branch currents at each node in the circuit to be simulated. The Branch Constitutive Equations (BCE) and Kirchoff's Voltage Law (KVL) are used to relate the branch currents to branch voltages and the branch voltages to node voltages, respectively. However, the usefulness of the nodal approach derives primarily from the fact that for both linear and nonlinear (or piecewise-linear) circuit problems, nodal equations may be generated by little more than inspection.

Element	Mode	Branch Relation
Resistor	DC,TR,AC	$V = R*I$ $I = V/R = G*V$
Capacitor	DC TR AC	$V = ?, I = 0$ $I = dQ(V)/dt$ $= (dQ/dV)(dV/dt)$ $= Q'V'$ $= C(V)\,dV/dt$ $I = j\omega CV$
Inductor	DC TR AC	$V = 0, I = ?$ $V = d\phi(I)/dt$ $= (d\phi/dI)(dI/dt)$ $= \phi'I'$ $= L(I)\,dI/dt$ $= j\omega LI$
Mutual Inductor	DC TR AC	$V_1 = V_2 = 0$ $I_1 = ?, I_2 = ?$ $V_1 = L_{11}\,dI_1/dt + M\,dI_2/dt$ $V_2 = M\,dI_1/dt + L_{22}\,dI_2/dt$ $V_1 = j\omega L_{11}I_1 + j\omega MI_2$ $V_2 = j\omega MI_1 + j\omega L_{22}I_2$
Voltage Source	DC,TR,AC	$V = V_S$ $I = ?$
Current Source	DC,TR,AC	$V = ?$ $I = I_S$
Voltage-Controlled Voltage Source	DC,TR,AC	$V_S = A_V*V_C$ $I_S = ?$
Voltage-Controlled Current Source	DC,TR,AC	$V_S = ?$ $I_S = G_T*V_C$
Current-Controlled Voltage Source	DC,TR,AC	$V_S = R_T*I_C$ $I_S = ?$
Current-Controlled Current Source	DC,TR,AC	$V_S = ?$ $I_S = A_I*I_C$

TABLE 1.1   - Branch Constitutive Equations

As an example, consider the case of a linear circuit. The nodal equations are of the form

$$Y V = I \qquad (1.1)$$

where Y is the nodal admittance matrix, V is the vector of node voltages to be found and I is a vector representing independent source currents. Viewed as a system of equations, each equation represents a statement of KCL at a node. The term y_{ii} in Y represents the sum of the admittances of all the branches connected to node i; y_{ij} is the negative of the sum of the admittances of all the branches connecting node i and node j; and i_k is the sum of all source currents entering node k. Thus if a resistor of value R connects nodes 3 and 5, $G(=1/R)$ is added to y_{33} and y_{55} and subtracted from y_{35} and y_{53}. Further, if a current source of strength I is directed from node 2 to node 4, I is subtracted from i_2 and added to i_4. In terms of the matrix form of (1.1), the resistor and current source would appear as

$$
\begin{array}{c}
\varepsilon_1: \\
\varepsilon_2: \\
\varepsilon_3: \\
\varepsilon_4: \\
\varepsilon_5: \\
\end{array}
\begin{bmatrix}
\cdot & & \cdot \\
 & \cdot & & \cdot \\
\cdots\; G & \cdot & -G \cdots \\
 & \cdot & & \cdot \\
\cdots -G & \cdot & G \cdots \\
\cdot & & \cdot \\
\cdot & & \cdot \\
\end{bmatrix}
\begin{bmatrix}
V_1 \\ V_2 \\ V_3 \\ V_4 \\ V_5 \\ \cdot \\ \cdot
\end{bmatrix}
=
\begin{bmatrix}
\cdot \\ -I \\ \cdot \\ +I \\ \cdot \\ \cdot \\ \cdot
\end{bmatrix}
$$

with columns labeled $\cdots\; 3\;\; 4\;\; 5 \cdots$ and RHS.

It can be seen that the appearance of each of the two elements is characterized by a distinct pattern. Similarly, each of the remaining basic elements whose branch constitutive equations are given in Table 1.1 is characterized by a similar pattern. It is this property of the nodal formulation approach that makes it attractive.

In slightly more abstract terms, the pattern for a resistor can be described as

	$V+$	$V-$	RHS
$\varepsilon+$	$1/R$	$-1/R$	
$\varepsilon-$	$-1/R$	$1/R$	

where $V+$ represents the column in Y corresponding to the positive reference node of R while $V-$ represents the column corresponding to the negative reference node. $\varepsilon+$ represents the KCL equation (row in Y) corresponding to the positive reference node while $\varepsilon-$ represents the KCL equation corresponding to the negative reference node. RHS represents a contribution to the right hand side (current vector) of (1.1). Similarly, the pattern for the current source can be represented as

	V_F	V_T	RHS
ε_F			$-I$
ε_T			$+I$

where current is directed from node F through the source to node T.

The patterns for those elements commonly allowed in nodal based programs are summarized in Table 1.2. Several comments should be made. First, for quiescent dc, capacitors behave as open circuits, hence, their contribution to Y is zero for dc. Conversely, inductors behave as short circuits for quiescent dc. This fact implies that the nodes to which an inductor is connected should be shorted together. Practically, it is more convenient to assume that inductors include a small series resistance of 1 ohm; hence, their contribution to Y is a self admittance of 1 mho for dc. Finally, for transient analysis it will later be shown that reactive elements (capacitors, inductors and mutual inductors) are modeled at each point in time as an equivalent conductance and current source in parallel.

Clearly, one very important circuit element, the voltage source, has been omitted from Table 1.2. The reason for its omission is that it is characterized by an infinite admittance as can be deduced from Table 1.1. Consequently, voltage sources in nodal analysis are treated as a special case. Two approaches have been used. The first is to insist that every voltage source appear in series with a resistor. This condition allows the transformation of the source to a Norton equivalent current source as illustrated in Figure 1.1.

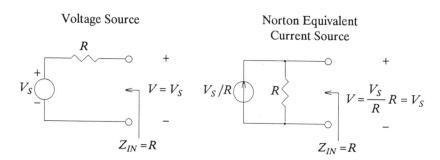

Figure 1.1 - Conversion of voltage sources to their Norton Equivalent Current Sources

The second approach is a slight generalization of the first but does not depend upon a series resistor. It is most easily introduced in terms of grounded voltage sources. The nodal equations are first assembled in terms of all elements included in Table 1.2. It is assumed that positive voltage source nodes are ordered sequentially before other nodes are considered. Columns of the admittance matrix corresponding to grounded voltage source nodes are then multiplied by the value of the source and

**Basic
Element**

**DC
Analysis**

Resistor

	$V+$	$V-$	RHS
$\varepsilon+$	$1/R$	$-1/R$	
$\varepsilon-$	$-1/R$	$1/R$	

Capacitor

	$V+$	$V-$	RHS
$\varepsilon+$	0	0	
$\varepsilon-$	0	0	

Inductor

	$V+$	$V-$	RHS
$\varepsilon+$	1	-1	
$\varepsilon-$	-1	1	

**Current
Source**

	V_F	V_T	RHS
ε_F			$-I$
ε_T			$+I$

**Voltage
Controlled
Current
Source**

	V_F	V_T	$V+$	$V-$	RHS
ε_F			G_T	$-G_T$	
ε_T			$-G_T$	G_T	
$\varepsilon+$					
$\varepsilon-$					

**Mutual
Inductor**

	V_1^+	V_1^-	V_2^+	V_2^-	RHS
ε_1^+	1	-1			
ε_1^-	-1	1			
ε_2^+			1	-1	
ε_2^-			-1	1	

TABLE 1.2a - Nodal Admittance Matrix Element Patterns for DC Analysis

Basic Element		Transient Analysis		

Resistor

	$V+$	$V-$	RHS
$\varepsilon+$	$1/R$	$-1/R$	
$\varepsilon-$	$-1/R$	$1/R$	

Capacitor

	$V+$	$V-$	RHS
$\varepsilon+$	G_C	$-G_C$	$-I_C$
$\varepsilon-$	$-G_C$	G_C	$+I_C$

Inductor

	$V+$	$V-$	RHS
$\varepsilon+$	G_L	$-G_L$	$-I_L$
$\varepsilon-$	$-G_L$	G_L	$+I_L$

Current Source

	V_F	V_T	RHS
ε_F			$-I$
ε_T			$+I$

Voltage Controlled Current Source

	V_F	V_T	$V+$	$V-$	RHS
ε_F			G_T	$-G_T$	
ε_T			$-G_T$	G_T	
$\varepsilon+$					
$\varepsilon-$					

Mutual Inductor

	V_1^+	V_1^-	V_2^+	V_2^-	RHS
ε_1^+	G_{11}	$-G_{11}$	G_M	$-G_M$	$-I_{11}-I_M$
ε_1^-	$-G_{11}$	G_{11}	$-G_M$	G_M	$I_{11}+I_M$
ε_2^+	G_M	$-G_M$	G_{22}	$-G_{22}$	$-I_{22}-I_M$
ε_2^-	$-G_M$	G_M	$-G_{22}$	G_{22}	$I_{22}+I_M$

TABLE 1.2b - Nodal Admittance Matrix Element Patterns for Transient Analysis

Basic Element

AC Analysis

Resistor

	$V+$	$V-$	RHS
$\varepsilon+$	$1/R$	$-1/R$	
$\varepsilon-$	$-1/R$	$1/R$	

Capacitor

	$V+$	$V-$	RHS
$\varepsilon+$	$j\omega C$	$-j\omega C$	
$\varepsilon-$	$-j\omega C$	$j\omega C$	

Inductor

	$V+$	$V-$	RHS
$\varepsilon+$	$\Gamma/j\omega$	$-\Gamma/j\omega$	
$\varepsilon-$	$-\Gamma/j\omega$	$\Gamma/j\omega$	

Current Source

	V_F	V_T	RHS
ε_F			$-I$
ε_T			$+I$

Voltage Controlled Current Source

	V_F	V_T	$V+$	$V-$	RHS
ε_F			G_T	$-G_T$	
ε_T			$-G_T$	G_T	
$\varepsilon+$					
$\varepsilon-$					

Mutual Inductor

	V_1^+	V_1^-	V_2^+	V_2^-	RHS
ε_1^+	$\Gamma_{11}/j\omega$	$-\Gamma_{11}/j\omega$	$\Gamma_M/j\omega$	$-\Gamma_M/j\omega$	
ε_1^-	$-\Gamma_{11}/j\omega$	$\Gamma_{11}/j\omega$	$-\Gamma_M/j\omega$	$\Gamma_M/j\omega$	
ε_2^+	$\Gamma_M/j\omega$	$-\Gamma_M/j\omega$	$\Gamma_{22}/j\omega$	$-\Gamma_{22}/j\omega$	
ε_2^-	$-\Gamma_M/j\omega$	$\Gamma_M/j\omega$	$-\Gamma_{22}/j\omega$	$\Gamma_{22}/j\omega$	

TABLE 1.2c - Nodal Admittance Matrix Element Patterns for AC Small-Signal Analysis

the result is subtracted from both sides of (1.1). This procedure can be illustrated symbolically as follows: Let the subscript V denote quantities associated with voltage source nodes while the subscript N denotes quantities associated with the remaining nodes. Then Y, V and I are partitioned such that (1.1) may be written

$$Y_V V_V + Y_N V_N = I_N - I_{SV} \qquad (1.2)$$

where I_{SV} is the as yet unknown value of the current flowing through the voltage source from the positive node to ground. The subtraction of $Y_V V_V$ from and the addition of I_{SV} to both sides of (1.2) results in

$$+I_{SV} + Y_N V_N = I_N - Y_V V_V \qquad (1.3)$$

Pictorially in matrix form, (1.2) and (1.3) may be presented as

$$\begin{bmatrix} Y_V & | & Y_N \end{bmatrix} \begin{bmatrix} V_V \\ -- \\ V_N \end{bmatrix} = \begin{bmatrix} I_N \end{bmatrix} - \begin{bmatrix} I_{SV} \\ -- \\ 0 \end{bmatrix} \qquad (1.4)$$

and

$$\begin{bmatrix} +1 & | \\ -- & | & Y_N \\ 0 & | \end{bmatrix} \begin{bmatrix} I_{SV} \\ -- \\ V_N \end{bmatrix} = \begin{bmatrix} I_N \end{bmatrix} - \begin{bmatrix} Y_V & | & 0 \end{bmatrix} \begin{bmatrix} V_V \\ -- \\ 0 \end{bmatrix} \qquad (1.5)$$

respectively where, by introducing the identity sub-matrix, I_{SV} has been incorporated into the vector of unknowns.

If the voltage source current I_{SV} is required, it may be solved for as an unknown. Otherwise, the equation representing the sum of the currents at the source node may be considered extraneous and dropped, thereby reducing the number of unknowns.

The extension of the above technique to floating voltage sources can similarly be illustrated with only minor changes in notation. The key observation required is that floating voltage sources constrain certain node voltages with respect to other node voltages according to the equation

$$V_{PV} = V_{NV} + V_{SV} \qquad (1.6)$$

where V_{PV} now represents the node voltages at positive voltage source nodes, V_{NV} represents the node voltages at negative voltage source nodes and finally, V_{SV} represents the values of the voltage sources. If Y is further partitioned, equations (1.2) and (1.3) become

$$Y_{PV} V_{PV} + Y_N V_N + Y_{NV} V_{NV} = I_N - I_{SV} \qquad (1.7)$$

and

$$I_{SV} + Y_N V_N + (Y_{NV} + Y_{PV}) V_{NV} = I_N - Y_{PV} V_{SV} \qquad (1.8)$$

respectively. Note that (1.6) has been used to eliminate V_{PV} from (1.7) in order to arrive at (1.8). Again, pictorially the process can be represented as follows:

$$\left[V_{PV}\right] = \left[V_{NV}\right] + \left[V_{SV}\right] \qquad (1.9)$$

$$
\begin{bmatrix}
 & | & | \\
 & | & | \\
Y_{PV} & | \ Y_N \ | & Y_{NV} \\
 & | & | \\
 & | & |
\end{bmatrix}
\begin{bmatrix}
V_{SV} \\
-- \\
V_N \\
-- \\
V_{NV}
\end{bmatrix}
=
\begin{bmatrix}
\ \\
I_N \\
\
\end{bmatrix}
-
\begin{bmatrix}
+1 \\
-- \\
0 \\
-- \\
A
\end{bmatrix}
\left[I_{SV}\right]
\qquad (1.10)
$$

$$
\begin{bmatrix}
+1 & | & | \\
-- & | & | \\
0 & | \ Y_N \ | & Y_{NV} \oplus Y_{PV} \\
-- & | & | \\
A & | & |
\end{bmatrix}
\begin{bmatrix}
I_{SV} \\
-- \\
V_N \\
-- \\
V_{NV}
\end{bmatrix}
=
\begin{bmatrix}
\ \\
I_N \\
\
\end{bmatrix}
-
\begin{bmatrix}
\ \\
Y_V \\
\
\end{bmatrix}
\left[V_{SV}\right]
\qquad (1.11)
$$

where $\oplus$ indicates the addition of columns corresponding to the positive source node to columns corresponding to the negative source node and the sub-matrix A, consisting of -1's and 0's, reflects the flow of the unknown source currents toward the negative voltage source nodes. Note that A can be reduced to zero by adding rows corresponding to positive voltage source nodes to those of negative source nodes. Clearly, if all voltage sources are grounded, $V_{NV} = 0$ such that (1.11) reduces to (1.5).

The practical problem of treating floating voltage sources as just illustrated arises when two voltages sources are placed in series. This situation as well as others involving the remaining types of controlled sources may be handled by additional column and row operations. Modified nodal analysis, to be considered shortly, provides a more systematic framework for treating these issues.

Before leaving the topic of nodal equation formulation, it is important to consider the nature of the reference voltage with respect to which each node voltage is measured or computed. While in a physical circuit the choice of a ground or reference node may not be left to one's choosing, computer simulation affords the user complete freedom of choice. One node may serve as a datum or reference node as well as any other. As a consequence most simulation programs formulate a node equation for the datum node as well as all non-datum nodes. This procedure avoids the necessity to treat grounded elements as special cases, each of which would have patterns which were subsets of those in Table 1.2. The result of this procedure is an augmented nodal admittance matrix denoted as Y_A which is generally referred to as the indefinite admittance matrix. The normal or definite admittance matrix Y of (1.1) can always be obtained from Y_A by deleting the row and column of Y_A corresponding to the datum node.

The indefinite admittance matrix Y_A has an important property which is that the sum of all terms in any row or column equals zero. This property becomes obvious when the patterns of Table 1.2 are examined and will be found useful when the attainment of accurate solutions to the simultaneous linear nodal equations is considered later.

1.3 Modified Nodal Analysis

As previously brought out, the conceptual short-coming of nodal analysis in terms of computer implementation is the cumbersome manner in which voltage sources and current-controlled sources are handled. Though not specifically brought out, a third problem is the direct evaluation of branch currents as output quantities. A systematic and straight-forward approach intended to deal with these problems while yet retaining the simplicity and other advantages of nodal analysis was implemented by Ho, et. al. of IBM [E 12]. The formulation, referred to as modified nodal analysis, will now be described.

For the purpose of illustration, again assume that the circuit to be simulated is linear such that equations may be written in matrix form. Written in Ho's notation, the matrix circuit equation becomes

$$
\begin{bmatrix} Y_R & B \\ C & D \end{bmatrix} \begin{bmatrix} V \\ I \end{bmatrix} = \begin{bmatrix} J \\ F \end{bmatrix}
\tag{1.12}
$$

where Y_R is a reduced form of the nodal admittance matrix Y of (1.1) containing only those elements whose patterns are given in Table 1.2 and whose branch currents are not required as output quantities. The partition B represents the contributions to KCL at each node of the additional output or controlling current variables, I. The partitions C and D when multiplied by V and I respectively and equated to F represent the branch constitutive equations of those elements which are current-controlled, whose currents are unknown, or whose currents represent output quantities. Finally, J represents independent source currents while F represents independent source voltages and/or currents where the latter are also output quantities.

If the above description is somewhat confusing, a simple example should help to clarify the situation. An elementary resistive circuit is shown in Figure 1.2.

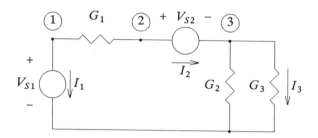

Figure 1.2 - Elementary resistive circuit with floating voltage source and current output

The modified nodal equations for this circuit can be written in matrix forms as

$$
\begin{array}{c}
\varepsilon_1: \\
\varepsilon_2: \\
\varepsilon_3: \\
\\
\varepsilon_4: \\
\varepsilon_5: \\
\varepsilon_6:
\end{array}
\left[
\begin{array}{ccc|ccc}
G_1 & -G_1 & 0 & 1 & 0 & 0 \\
-G_1 & G_1 & 0 & 0 & 1 & 0 \\
0 & 0 & G_2 & 0 & -1 & 1 \\
\hline
-1 & 0 & 0 & 0 & 0 & 0 \\
0 & -1 & +1 & 0 & 0 & 0 \\
0 & 0 & G_3 & 0 & 0 & -1
\end{array}
\right]
\left[
\begin{array}{c}
V_1 \\
V_2 \\
V_3 \\
\hline
I_1 \\
I_2 \\
I_3
\end{array}
\right]
=
\left[
\begin{array}{c}
0 \\
0 \\
0 \\
\hline
-V_{S1} \\
-V_{S2} \\
0
\end{array}
\right]
\qquad (1.13)
$$

Equations ε_1 ε_2 and ε_3 can be seen to be the nodal equations for nodes 1, 2 and 3 respectively. Note that B contains the direct contributions of the auxilliary current variables I_1, I_2 and I_3 while Y_R contains conventional admittance terms. Equations ε_4 and ε_5 relate the node voltages of V to the source voltages of F. Finally, equation ε_6 represents the branch constitutive equation for the conductance G_3 whose current is an output variable. Note that the node equation corresponding to the ground node has already been omitted.

In general, the unknown variables established in the modified nodal approach consist of

1)   Node voltages

2)   Voltage source currents

3)   Output currents

4)   Controlling source currents

This choice results in a linear, independent system of equations in which the nonzero structure of the coefficient matrix on the left-hand side of (1.12) is symmetric in the absence of controlled sources. As with conventional nodal analysis, the equations can be written directly by inspection. For modified nodal analysis, the nonzero patterns for the basic circuit elements of Table 1.1 are given in Table 1.3 for the case of transient analysis. The extension of dc and ac analyses follows as in Table 1.2. Note that in Table 1.3, two cases are distinguished -- the normal case resulting in terms of Y_R and the case where the branch current is an output variable. It should be pointed out that these patterns differ slightly in terms of sign convention from those given by Ho. The change was made in order to preserve consistency as much as possible.

As a final observation before proceeding on to the sparse tableau approach, note that a number of patterns in Table 1.3 are characterized by zeros for diagonal elements. For those already familiar with the application of Gaussian elimination methods in solving systems of linear nodal equations, this observation may be of concern. However, it will be pointed out in the next chapter that with care in the order of pivot selection, pivoting on the diagonal can be retained.

If the equations, (1.13), are considered by way of an example, it can be seen that for the existing ordering, the elimination of the -1 in the first column of equation ε_4 will result in the creation of a nonzero term at the diagonal position of equation ε_4. A similar nonzero term will be generated at the diagonal position of equation ε_5 when the -1 in the second column is eliminated.

1.4 Sparse Tableau Analysis

The final equation formulation approach to be considered in this section is the sparse tableau formulation of Hachtel, et. al. [E9]. While developed prior to the modified nodal formulation, the sparse tableau approach represents an extension of previously presented concepts to their basic, fundamental equations. That is, the set of unknowns is extended to include all branch currents, all branch voltages and all node voltages. A Kirchoff current law (KCL) equation is written in terms of branch currents for each node. A Kirchoff voltage law (KVL) equation relating a branch voltage to its node voltages is written for each branch. Finally, a branch constitutive

Current Not an Output Variable   **Current as Output Variable**

Resistor

	$V+$	$V-$	RHS
$\varepsilon+$	$1/R$	$-1/R$	
$\varepsilon-$	$-1/R$	$1/R$	

	$V+$	$V-$	I_R	RHS
$\varepsilon+$			$+1$	
$\varepsilon-$			-1	
ε_R	$1/R$	$-1/R$	-1	

Capacitor

	$V+$	$V-$	RHS
$\varepsilon+$	$\hat{G}_C$	$-\hat{G}_C$	$-\hat{I}_C$
$\varepsilon-$	$-\hat{G}_C$	$\hat{G}_C$	$+\hat{I}_C$

	$V+$	$V-$	I_C	RHS
$\varepsilon+$			$+1$	
$\varepsilon-$			-1	
ε_C	$\hat{G}_C$	$-\hat{G}_C$	-1	$-\hat{I}_C$

Inductor

	$V+$	$V-$	RHS
$\varepsilon+$	$\hat{G}_L$	$-\hat{G}_L$	$-\hat{I}_L$
$\varepsilon-$	$-\hat{G}_L$	$\hat{G}_L$	$+\hat{I}_L$

	$V+$	$V-$	I_L	RHS
$\varepsilon+$			$+1$	
$\varepsilon-$			-1	
ε_L	$\hat{G}_L$	$-\hat{G}_L$	-1	$-\hat{I}_L$

Current Source

	V_F	V_T	RHS
ε_F			$-I$
ε_T			$+I$

	V_F	V_T	I_I	RHS
ε_F			$+1$	
ε_T			-1	
ε_I			-1	$-I$

Voltage Source

	$V+$	$V-$	I_V	RHS
$\varepsilon+$			$+1$	
$\varepsilon-$			-1	
ε_V	-1	$+1$		$-V$

TABLE 1.3a - Basic Element Patterns for Modified Nodal Analysis (Transient Analysis Only)

Current Not an Output Variable   **Current as Output Variable**

Voltage Controlled Voltage Source

	V_{S+}	V_{S-}	V_{C+}	V_{C-}	I_{VS}
ε_{S+}					$+1$
ε_{S-}					-1
ε_{C+}					
ε_{C-}					
ε_{VS}	-1	$+1$	A_V	$-A_V$	

Voltage Controlled Current Source

	V_F	V_T	$V+$	$V-$
ε_F			G_T	$-G_T$
ε_T			$-G_T$	G_T
$\varepsilon+$				
$\varepsilon-$				

	V_{SF}	V_{ST}	V_{C+}	V_{C-}	I_{GT}
ε_{SF}					$+1$
ε_{ST}					-1
ε_{C+}					
ε_{C-}					
ε_{GT}			G_T	$-G_T$	-1

Current Controlled Voltage Source

	V_{S+}	V_{S-}	V_{CF}	V_{CT}	I_{RT}
ε_{S+}					
ε_{S-}					
ε_{CF}					$+1$
ε_{CT}					-1
ε_{RT}	-1	$+1$			R_T

Current Controlled Current Source (Resistive Branch)

	V_{SF}	V_{ST}	V_{CF}	V_{CT}	I_{IS}	I_{CS}
ε_{SF}					$+1$	
ε_{ST}					-1	
ε_{CF}						$+1$
ε_{CT}						-1
ε_{AI}					-1	A_I
ε_{CS}						-1

TABLE 1.3b - Basic Element Patterns for Modified Nodal Analysis (Transient Analysis Only)

equation (BCE) is written for each branch in terms of its branch voltage and current. Collectively, these equations may be written as

$$A\, I_B = 0 \qquad \text{(KCL)} \tag{1.14}$$

$$V_B = A^T V_N \qquad \text{(KVL)} \tag{1.15}$$

$$I_{BG} + G\, V_{BG} = I_S \qquad \text{(BCE)} \tag{1.16}$$

$$R\, I_{BR} + V_{BR} = V_S \qquad \text{(BCE)} \tag{1.17}$$

respectively where the branch currents I_B and the branch voltages V_B have been partitioned with respect to voltage and current defined branches such that

$$I_B = I_{BG} \,|\, I_{BR} \tag{1.18}$$

$$V_B = V_{BG} \,|\, V_{BR} \tag{1.19}$$

In (1.14)-(1.17), A is the node-to-branch incidence matrix defined such that

$$a_{ij} = \begin{cases} 0 & \text{if node } i \text{ is not connected branch } j \\ +1 & \text{if node } i \text{ is the positive node of branch } j \\ -1 & \text{if node } i \text{ is the negative node of branch } j \end{cases}$$

V_N is the vector of node voltages and G and R represent the branch constitutive equations of voltage and current-defined branches respectively while I_S and V_S are vectors of independent current and voltage sources respectively. Written in the form of a partitioned matrix equation as

$$\begin{bmatrix} 0 & A & 0 \\ A^T & 0 & -1 \\ 0 & 1\,|\,R\;G\,|\,1 \end{bmatrix} \begin{bmatrix} V_N \\ I_B \\ V_B \end{bmatrix} = \begin{bmatrix} 0 \\ 0 \\ I_S \\ V_S \end{bmatrix} \tag{1.20}$$

(1.14)-(1.17) represent the sparse tableau formulation. In expanded form, (1.20) becomes

$$\begin{bmatrix} 0 & A & 0 \\ A^T & 0 & -1 \\ 0 & 1\,|\,0\;G\,|\,0 \\ 0 & 0\,|\,R\;0\,|\,1 \end{bmatrix} \begin{bmatrix} V_N \\ I_{BG} \\ I_{BR} \\ V_{BG} \\ V_{BR} \end{bmatrix} = \begin{bmatrix} 0 \\ 0 \\ I_S \\ V_S \end{bmatrix} \tag{1.21}$$

The similarity between (1.21) and the patterns for the basic circuit elements given in Table 1.3 for the modified nodal approach should be apparent. Further, if (1.17) is multiplied by R^{-1}, it is easily seen from (1.21) that (1.14)-(1.17) may be combined to give

$$A \; G \mid R^{-1} A^T \; V_N = I_S + R^{-1} V_S \tag{1.22}$$

or

$$Y \; V_N = I_S \tag{1.23}$$

where

$$G \mid R^{-1} = \begin{bmatrix} G & 0 \\ 0 & R^{-1} \end{bmatrix} \tag{1.24}$$

such that

$$Y = A \; G \mid R^{-1} A^T \tag{1.25}$$

while in (1.23) it is assumed $V_S = 0$. It can thus be seen that nodal equations are directly derivable from the sparse tableau formulation.

One final observation should be made before considering the solution of systems of linear equations in the next chapter. It is usually contended that the argument in favor of the sparse tableau formulation is that while there are more equations per system, there are fewer nonzero terms per equation and consequently, the number of mathematical operations required to solve those equations can be less than the number required to solve smaller, more dense systems such as derived from nodal analysis. Support for this argument is based upon the fact that many, if not most, nonzero terms in (1.21) are $+1$ or -1, whose processing need only be done once and whose values need not be stored. Many other terms are constant resistances or conductances whose values do not change with time or operating point and so only need to be processed once per timepoint. This idea can be extended to linear charge and flux storage elements and finally to nonlinear elements which vary with every Newton iteration for each timepoint. Hachtel, *et. al.* made use of this observation in assigning each nonzero element a variability type or weighting according to the frequency with which it must be processed during equation solution. An attempt is then made to process first those elements which vary least frequently and to process last those elements which vary most frequently. It has been shown that this procedure results in a very efficient solution.

2. LINEAR EQUATION SOLUTION

2.1 Gaussian Elimination

In the previous chapter it was stated that the circuit equations for linear and nonlinear circuits could be reduced to a system of simultaneous linear algebraic equations of the form of (1.1), (1.2) or (1.20) for quiescent dc, small-signal ac or large-signal transient analyses. For nonlinear and time domain analyses, the method of reduction to linear algebraic equations has yet to be demonstrated. Nonetheless, in this chapter methods of solving systems of linear algebraic equations will be considered. As the methods apply to nodal analysis, modified nodal analysis and sparse tableau analysis equally well, a system of linear equations of the form

$$A X = B \tag{2.1}$$

will be assumed.

With the distinction in mind that the system of equation (2.1) involves only real variables in the dc or transient case and complex variables in the ac case, the solution of (2.1) can be written

$$X = A^{-1}B \tag{2.2}$$

where A^{-1} is the inverse of the matrix A. Computationally, the efficiency of this approach can be evaluated in terms of the number of long operations (multiplications and divisions) required. It can be shown that the number of long operations required to invert an $N \times N$ matrix is N^3, while the number of long operations required to compute its product with a vector of dimension N is N^2. Thus the total number of long operations required by the approach is

$$N^3 + N^2$$

The number of long operations required to obtain a solution to (2.1) can be reduced by a factor of 3 using Gaussian elimination.

The Gaussian elimination procedure consists of two steps. The first step is the conversion of the system (2.1) into an equivalent upper-triangular system, i.e., a system in which the new matrix $\hat{A}$ contains only zero-valued elements below the diagonal. The second step, which is referred to as back substitution, consists of solving the Nth equation containing only x_N for x_N, the N-1st equation for x_{N-1} in terms of x_N, etc. Gaussian elimination will be illustrated for the third-order system

$$
\begin{array}{l}
\varepsilon_1^{(0)}: \\
\varepsilon_2^{(0)}: \\
\varepsilon_3^{(0)}:
\end{array}
\begin{bmatrix}
a_{11}^{(0)} & a_{12}^{(0)} & a_{13}^{(0)} \\
a_{21}^{(0)} & a_{22}^{(0)} & a_{23}^{(0)} \\
a_{31}^{(0)} & a_{32}^{(0)} & a_{33}^{(0)}
\end{bmatrix}
\begin{bmatrix}
x_1 \\
x_2 \\
x_3
\end{bmatrix}
=
\begin{bmatrix}
b_1^{(0)} \\
b_2^{(0)} \\
b_3^{(0)}
\end{bmatrix}
\tag{2.3}
$$

where the superscript 0 indicates the initial system of equations. As indicated previously, the unknown x_1 is eliminated from equations $\varepsilon_2^{(0)}$ and $\varepsilon_3^{(0)}$ by subtracting $(a_{21}^{(0)}/a_{11}^{(0)})\varepsilon_1^{(0)}$ from $\varepsilon_2^{(0)}$ and $(a_{31}^{(0)}/a_{11}^{(0)})\varepsilon_1^{(0)})$ from $\varepsilon_3^{(0)}$ Symbolically the transformation can be represented by the equations

$$
\varepsilon_1^{(1)} = \varepsilon_1^{(0)}
$$

$$
\varepsilon_2^{(1)} = \varepsilon_2^{(0)} - (a_{21}^{(0)}/a_{11}^{(0)})\varepsilon_1^{(0)}
\tag{2.4}
$$

$$
\varepsilon_3^{(1)} = \varepsilon_3^{(0)} - (a_{31}^{(0)}/a_{11}^{(0)})\varepsilon_1^{(0)}
$$

which yields the system

$$
\begin{array}{l}
\varepsilon_1^{(1)}: \\
\varepsilon_2^{(1)}: \\
\varepsilon_3^{(1)}:
\end{array}
\begin{bmatrix}
a_{11}^{(0)} & a_{12}^{(0)} & a_{13}^{(0)} \\
0 & a_{22}^{(1)} & a_{23}^{(1)} \\
0 & a_{32}^{(1)} & a_{33}^{(1)}
\end{bmatrix}
\begin{bmatrix}
x_1 \\
x_2 \\
x_3
\end{bmatrix}
=
\begin{bmatrix}
b_1^{(0)} \\
b_2^{(1)} \\
b_3^{(1)}
\end{bmatrix}
\tag{2.5}
$$

Finally, the unknown x_2 is eliminated from $\varepsilon_3^{(1)}$ by subtracting $(a_{32}^{(1)}/a_{22}^{(1)})\varepsilon_2^{(1)}$ from $\varepsilon_3^{(1)}$ thus obtaining

$$
\varepsilon_1^{(2)} = \varepsilon_1^{(1)} = \varepsilon_1^{(0)}
$$

$$
\varepsilon_2^{(2)} = \varepsilon_2^{(1)}
\tag{2.6}
$$

$$
\varepsilon_3^{(2)} = \varepsilon_3^{(1)} - (a_{32}^{(1)}/a_{22}^{(1)})\varepsilon_2^{(1)}
$$

which yields the upper-triangularized system

$$
\begin{array}{l}
\varepsilon_1^{(2)}: \\
\varepsilon_2^{(2)}: \\
\varepsilon_3^{(2)}:
\end{array}
\begin{bmatrix}
a_{11}^{(0)} & a_{12}^{(0)} & a_{13}^{(0)} \\
0 & a_{22}^{(1)} & a_{23}^{(1)} \\
0 & 0 & a_{33}^{(2)}
\end{bmatrix}
\begin{bmatrix}
x_1 \\
x_2 \\
x_3
\end{bmatrix}
=
\begin{bmatrix}
b_1^{(0)} \\
b_2^{(1)} \\
b_3^{(2)}
\end{bmatrix}
\tag{2.7}
$$

This result completes the first step in the Gaussian elimination procedure. Back substitution is now performed to obtain the final solution as follows

$$x_3 = b_3^{(2)}/a_{33}^{(2)}$$

$$x_2 = \left(b_2^{(1)} - a_{23}^{(1)}x_3\right)/a_{22}^{(1)} \tag{2.8}$$

$$x_1 = \left(b_1^{(0)} - a_{13}^{(0)}x_3 - a_{12}^{(0)}x_2\right)/a_{11}^{(0)}$$

If the right-hand-side vector B is appended to the matrix A as an $N+1$st column such that $a_{1,N+1}^{(0)} = b_1^{(0)}$, $a_{2,N+1}^{(0)} = b_2^{(0)}$, etc., the above procedure can be summarized in equation form as

$$a_{ij} = a_{ij} - \frac{a_{in}}{a_{nn}}a_{nj} \quad \left.\begin{array}{l} i = n+1, N \\ j = n, N+1 \end{array}\right\} \quad n = 1, N \tag{2.9}$$

for the triangularization at the N th step and

$$x_n = \left(a_{n,N+1} - \sum_{j=n+1}^{N} a_{nj}x_j\right)/a_{nn} \quad n = N, 1 \tag{2.10}$$

for the back substitution at the N th step. Careful enumeration shows that for an N th-order system, the number of long operations required by Gaussian elimination is

$$\frac{N^3}{3} + N^2 - \frac{N}{3}$$

A variation of the above procedure attributed to Jordan involves the elimination of x_2 from ε_1 at the second step, the elimination of x_3 from ε_1 and ε_2 at the third step, etc. such that the final reduced system is diagonalized rather than merely triangularized. This variation in effect performs back substitution concurrently with the reduction. Thus for the third order example (2.3) considered previously, the first step as described in (2.4) and (2.5) remains the same while the second step becomes

$$\varepsilon_1^{(2)} = \varepsilon_1^{(1)} - (a_{12}^{(1)}/a_{22}^{(1)})\varepsilon_2^{(1)}$$

$$\varepsilon_2^{(2)} = \varepsilon_2^{(1)} \tag{2.11}$$

$$\varepsilon_3^{(2)} = \varepsilon_3^{(1)} - (a_{32}^{(1)}/a_{22}^{(1)})\varepsilon_2^{(1)}$$

which yields

$$
\begin{array}{l}
\varepsilon_1^{(2)}: \\
\varepsilon_2^{(2)}: \\
\varepsilon_3^{(2)}:
\end{array}
\begin{bmatrix}
a_{11}^{(0)} & 0 & a_{13}^{(2)} \\
0 & a_{22}^{(1)} & a_{23}^{(2)} \\
0 & 0 & a_{33}^{(2)}
\end{bmatrix}
\begin{bmatrix}
x_1 \\ x_2 \\ x_3
\end{bmatrix}
=
\begin{bmatrix}
b_1^{(2)} \\ b_2^{(2)} \\ b_3^{(2)}
\end{bmatrix}
\tag{2.12}
$$

The third step (in effect a part of back substitution) becomes

$$
\varepsilon_1^{(3)} = \varepsilon_1^{(2)} - (a_{13}^{(2)}/a_{33}^{(2)})\varepsilon_3^{(2)}
$$

$$
\varepsilon_2^{(3)} = \varepsilon_2^{(2)} - (a_{23}^{(2)}/a_{33}^{(2)})\varepsilon_3^{(2)}
\tag{2.13}
$$

$$
\varepsilon_3^{(3)} = \varepsilon_3^{(2)}
$$

which yields the diagonalized system

$$
\begin{array}{l}
\varepsilon_1^{(3)}: \\
\varepsilon_2^{(3)}: \\
\varepsilon_3^{(3)}:
\end{array}
\begin{bmatrix}
a_{11}^{(0)} & 0 & 0 \\
0 & a_{22}^{(1)} & 0 \\
0 & 0 & a_{33}^{(2)}
\end{bmatrix}
\begin{bmatrix}
x_1 \\ x_2 \\ x_3
\end{bmatrix}
=
\begin{bmatrix}
b_1^{(3)} \\ b_2^{(3)} \\ b_3^{(3)}
\end{bmatrix}
\tag{2.14}
$$

Finally, the solution is obtained as

$$
x_1 = b_1^{(3)}/a_{11}^{(0)}
$$

$$
x_2 = b_2^{(3)}/a_{22}^{(1)}
\tag{2.15}
$$

$$
x_3 = b_3^{(3)}/a_{33}^{(2)}
$$

If, as before, the right-hand-side vector B is appended to A as an $N+1$st column, the above Gauss-Jordan procedure can be summarized in equation form as

$$
a_{ij} = a_{ij} - \frac{a_{in}}{a_{nn}} a_{nj}
\left.
\begin{array}{l}
i = 1, N \\
\ne n \\
j = n, N+1
\end{array}
\right\}
\quad n = 1, N
\tag{2.16}
$$

for the diagonalization at the Nth step and

$$
x_n = a_{n,N+1}/a_{nn} \quad n = 1, N
\tag{2.17}
$$

for the solution computation at the nth step. Enumeration shows that for an Nth order system, the number of long operations required by Gauss-Jordan elimination is

$$\frac{N^3 - N + 3}{2}$$

Thus, Gauss-Jordan represents a slightly less efficient solution procedure than Gaussian elimination.

2.2 LU Transformation

A more useful modification of Gaussian elimination where more than one source or right-hand-side vector is to be considered or where sensitivity calculations require the solution of the adjoint network (to be described later) is the LU transformation. This procedure consists of partitioning A into an upper-triangular matrix U and a lower-triangular matrix L, usually with ones on the diagonal, such that

$$A = L U \tag{2.18}$$

(The reduction of A into L and U will be considered shortly.) The resulting system

$$L U X = B \tag{2.19}$$

is solved in two stages. First

$$U X = L^{-1}B = \hat{B} \tag{2.20}$$

and secondly,

$$X = U^{-1}\hat{B} \tag{2.21}$$

In this case since both L and U are triangular, L^{-1} and U^{-1} are trivial to compute. Note that (2.21) is the same back substitution performed during Gaussian elimination while (2.20), which in effect solves for $\hat{b}_1$, $\hat{b}_2$ in terms of $\hat{b}_1$, etc. in a similar fashion, is referred to as forward substitution. Thus, the LU transformation method requires three steps:

1) Reduction of A to L and U.

2) Forward substitution with L.

3) Backward substitution with U.

Previously it was shown that Gaussian elimination resulted in the formation of an upper-triangular system of equations as illustrated for a third-order system in (2.7). It is reasonable to ask if a lower-triangular matrix L can be found which satisfies (2.18) when U is taken to be the triangularized matrix derived via Gaussian elimination. That is, for the earlier third-order example assume

$$
U = \begin{bmatrix} a_{11}^{(0)} & a_{12}^{(0)} & a_{13}^{(0)} \\ 0 & a_{22}^{(1)} & a_{23}^{(1)} \\ 0 & 0 & a_{33}^{(2)} \end{bmatrix}
\tag{2.22}
$$

It is not difficult to show that a lower-triangular matrix L can be found and in fact is a by-product of the procedure. For the example, L is given by

$$
L = \begin{bmatrix} 1 & 0 & 0 \\ \dfrac{a_{21}^{(0)}}{a_{11}^{(0)}} & 1 & 0 \\ \dfrac{a_{31}^{(0)}}{a_{11}^{(0)}} & \dfrac{a_{32}^{(1)}}{a_{22}^{(1)}} & 1 \end{bmatrix}
\tag{2.23}
$$

The elements in L are seen to be the coefficients in (2.4) and (2.6).

A convenient aspect of the preceding procedure is that since the diagonal elements of L are known to be ones, L and U can share the same memory locations originally assigned to A. Note further that the long operations count is the same as for Gaussian elimination. Once L and U have been computed, (2.20) and (2.21) can be applied repeatedly for different right-hand-side vectors B. For M such vectors, the total long operations count becomes

$$
\frac{N^3}{3} + M N_2 - \frac{N}{3}
$$

As with Gaussian elimination, the LU transformation procedure can be summarized in equation form as

$$
\left.
\begin{array}{ll}
l_{in} = a_{in}^{(n-1)}/a_{nn}^{(n-1)} & \quad i = n+1, N \\[2ex]
u_{nj} = a_{nj}^{(n-1)} & \quad j = n, N \\[2ex]
a_{ij}^{(n)} = a_{ij}^{(n-1)} - l_{in}\, u_{nj} & \quad \begin{array}{l} i = n+1, N \\ j = n+1, N \end{array}
\end{array}
\right\} \quad n = 1, N
\tag{2.24}
$$

for the LU factorization at the n th step,

$$\hat{b}_n = b_n - \sum_{j=1}^{n-1} l_{nj} \hat{b}_j \qquad n = 1, N \qquad (2.25)$$

for the forward substitution at the nth step, and

$$x_n = \left(\hat{b}_n - \sum_{j=n+1}^{N} u_{nj} x_j \right) / u_{nn} \qquad n = N, 1 \qquad (2.26)$$

for the backward substitution at the nth step. These steps are illustrated in Figure 2.1.

2.3 LU Transformation Variations

Before the advent of high-speed digital computers, the primary means of solving systems of linear equations was by hand with the aid of a desk calculator. As a consequence, a great deal of attention was given to the order, arrangement and layout of linear systems during the solution process in order to eliminate the recording of intermediate systems such as (2.3),(2.5) and (2.7). The final results obtained by these desk top methods, including operations count, is identical to those already considered. They differ only in the order in which the various elements of U and L are calculated.

The first method to be considered is that of Doolittle in which elements are calculated by rows in the order

$$u_{11} \ u_{12} \ u_{13} \ u_{14} \ \cdots \ u_{1N}$$

$$l_{21} \ \cdot$$

$$u_{22} \ u_{23} \ u_{24} \ \cdots \ u_{2N}$$

$$l_{31} \ l_{32}$$

$$u_{33} \ u_{34} \ \cdots \ u_{3N}$$

$$l_{41} \ l_{42} \ l_{43}$$

$$\cdot$$

$$\cdot$$

$$\cdot$$

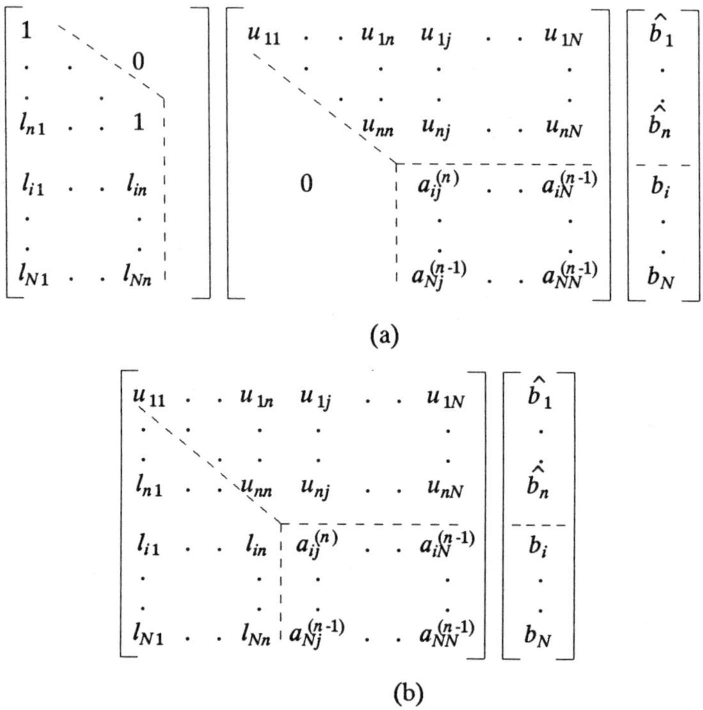

$$
\begin{bmatrix}
1 & & \diagdown & \\
 & \diagdown & & 0 \\
\vdots & \ddots & & \\
l_{n1} & \cdots & 1 & \\
l_{i1} & \cdots & l_{in} & \\
\vdots & & \vdots & \\
l_{N1} & \cdots & l_{Nn} &
\end{bmatrix}
\begin{bmatrix}
u_{11} & \cdot & \cdot & u_{1n} & u_{1j} & \cdot & \cdot & u_{1N} \\
 & \diagdown & & \cdot & \cdot & & & \cdot \\
 & & \diagdown & \cdot & \cdot & & & \cdot \\
 & & & u_{nn} & u_{nj} & \cdot & \cdot & u_{nN} \\
 & 0 & & a_{ij}^{(n)} & \cdot & \cdot & a_{iN}^{(n-1)} \\
 & & & a_{Nj}^{(n-1)} & \cdot & \cdot & a_{NN}^{(n-1)}
\end{bmatrix}
\begin{bmatrix}
\hat{b}_1 \\
\cdot \\
\hat{b}_n \\
b_i \\
\cdot \\
b_N
\end{bmatrix}
$$

(a)

$$
\begin{bmatrix}
u_{11} & \cdot & \cdot & u_{1n} & u_{1j} & \cdot & \cdot & u_{1N} \\
 & \diagdown & & \cdot & \cdot & & & \cdot \\
l_{n1} & \cdot & \cdot & u_{nn} & u_{nj} & \cdot & \cdot & u_{nN} \\
l_{i1} & \cdot & \cdot & l_{in} & a_{ij}^{(n)} & \cdot & \cdot & a_{iN}^{(n-1)} \\
 & & & \cdot & \cdot & & & \\
l_{N1} & \cdot & \cdot & l_{Nn} & a_{Nj}^{(n-1)} & \cdot & \cdot & a_{NN}^{(n-1)}
\end{bmatrix}
\begin{bmatrix}
\hat{b}_1 \\
\cdot \\
\hat{b}_n \\
b_i \\
\cdot \\
b_N
\end{bmatrix}
$$

(b)

$$
l_{in} = a_{in}^{(n-1)} / a_{nn}^{(n-1)} \quad u_{nj} = a_{nj}^{(n-1)} \quad a_{ij}^{(n)} = a_{ij}^{(n-1)} - l_{in} u_{nj} \quad \hat{b}_n = b_n - \sum_{j=1}^{n-1} l_{nj} \hat{b}_j
$$

Figure 2.1 - (a) Illustration of n th step in LU transformation as described by (2.24) - (2.26) and (b) of merged array element storage

The procedure can be summarized in equation form as

$$\left. \begin{array}{ll} u_{nj} = a_{nj} - \displaystyle\sum_{k=1}^{n-1} l_{nk} u_{kj} & j = n, N \\[4mm] l_{nj} = (a_{nj} - \displaystyle\sum_{k=1}^{n-1} l_{nk} u_{kj})/u_{nn} & j = 1, n-1 \end{array} \right\} \quad n = 1, N \qquad (2.27)$$

for the LU factorization at the n th step and is illustrated in Figure 2.2. The forward and backward substitution steps are the same as for the previous LU transformation procedure. Note that the elements of both U and L are computed in the form of inner products. This feature facilitates the use of extended precision for more accurate accumulation of partial sums and thus more accurate calculations in general. Finally, note that the elements of L and U are computed by rows, a feature which facilitated hand calculation.

Although Doolittle's method is often attributed to Crout, the two differ in that in Crout's method, ones are constructed on the diagonal of U rather than L, again to facilitate hand calculation. The elements of U and L are computed in the same order as for Doolittle's method. However, the nth step in the factorization is now described in equation form by

$$\left. \begin{array}{ll} u_{nn} = a_{nn} - \displaystyle\sum_{k=1}^{n-1} l_{nk} u_{kj} & \\[4mm] u_{nj} = (a_{nj} - \displaystyle\sum_{k=1}^{n-1} l_{nk} u_{kj})/u_{nn} & j = n+1, N \\[4mm] l_{nj} = a_{nj} - \displaystyle\sum_{k=1}^{j-1} l_{nk} u_{kj} & j = 1, n-1 \end{array} \right\} \quad n = 1, N \qquad (2.28)$$

Again the elements of U and L are computed by rows in the form of inner products.

It has been recognized more recently that for non-sparse matrices, digital computers store, and consequently process faster, vector and matrix data in column order. Further, where accuracy requirements dictate the use of partial pivoting, a technique requiring row interchanges which maximize the magnitude of the pivot or diagonal element at each step of the reduction, the processing of elements of U and L by column order is more advantageous. This choice results from the fact that the element of maximum magnitude on or below the diagonal in the affected column cannot be ascertained until all operations required by previous elimination steps are performed on the present column. The order in which elements of U and L are

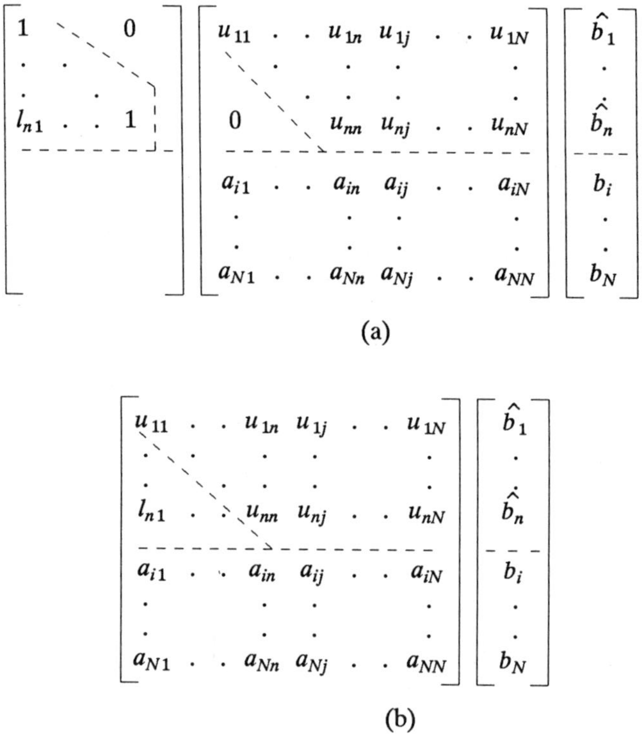

$$
\begin{bmatrix}
1 & \diagdown & & & 0 \\
\cdot & \cdot & & & \\
\cdot & & \cdot & & \\
l_{n1} & \cdot & \cdot & 1 & \\
\end{bmatrix}
\begin{bmatrix}
u_{11} & \cdot & \cdot & u_{1n} & u_{1j} & \cdot & \cdot & u_{1N} \\
 & \cdot & & \cdot & \cdot & & & \cdot \\
 & & \cdot & \cdot & \cdot & & & \cdot \\
0 & & & u_{nn} & u_{nj} & \cdot & \cdot & u_{nN} \\
\hline
a_{i1} & \cdot & \cdot & a_{in} & a_{ij} & \cdot & \cdot & a_{iN} \\
\cdot & & & \cdot & \cdot & & & \cdot \\
\cdot & & & \cdot & \cdot & & & \cdot \\
a_{N1} & \cdot & \cdot & a_{Nn} & a_{Nj} & \cdot & \cdot & a_{NN} \\
\end{bmatrix}
\begin{bmatrix}
\hat{b}_1 \\
\cdot \\
\cdot \\
\hat{b}_n \\
\hline
b_i \\
\cdot \\
\cdot \\
b_N \\
\end{bmatrix}
$$

(a)

$$
\begin{bmatrix}
u_{11} & \cdot & \cdot & u_{1n} & u_{1j} & \cdot & \cdot & u_{1N} \\
\cdot & \cdot & & \cdot & \cdot & & & \cdot \\
\cdot & & \cdot & \cdot & \cdot & & & \cdot \\
l_{n1} & \cdot & \cdot & u_{nn} & u_{nj} & \cdot & \cdot & u_{nN} \\
\hline
a_{i1} & \cdot & \cdot & a_{in} & a_{ij} & \cdot & \cdot & a_{iN} \\
\cdot & & & \cdot & \cdot & & & \cdot \\
\cdot & & & \cdot & \cdot & & & \cdot \\
a_{N1} & \cdot & \cdot & a_{Nn} & a_{Nj} & \cdot & \cdot & a_{NN} \\
\end{bmatrix}
\begin{bmatrix}
\hat{b}_1 \\
\cdot \\
\cdot \\
\hat{b}_n \\
\hline
b_i \\
\cdot \\
\cdot \\
b_N \\
\end{bmatrix}
$$

(b)

$$
l_{nj} = \left(a_{nj} - \sum_{k=1}^{n-1} l_{nk} u_{kj}\right)/u_{nn} \qquad u_{nj} = a_{nj} - \sum_{k=1}^{n-1} l_{nk} u_{kj} \qquad \hat{b}_n = b_n - \sum_{j=1}^{n-1} l_{nj} \hat{b}_j
$$

Figure 2.2 - (a) Illustration of n th step in Doolittle variation of LU factorization and (b) of merged array element storage

computed is

$$u_{11}$$

$$l_{21}\ l_{31}\ l_{41}\ l_{51}\ \cdots\ l_{N1}$$

$$u_{12}\ u_{22}$$

$$l_{32}\ l_{42}\ l_{52}\ \cdots\ l_{N2}$$

$$u_{13}\ u_{23}\ u_{33}$$

$$l_{43}\ l_{53}\ \cdots\ l_{N3}$$

$$.$$
$$.$$
$$.$$

For reduction without partial pivoting, the nth step in the reduction is given by

$$
\left.
\begin{aligned}
u_{in} &= a_{in} - \sum_{k=1}^{i-1} l_{ik} u_{kn} & i &= 1, n \\[2ex]
l_{in} &= \left(a_{in} - \sum_{k=1}^{n-1} l_{ik} u_{kn}\right)/u_{nn} & i &= n+1, N
\end{aligned}
\right\} \quad n = 1, N
\qquad (2.29)
$$

while for reduction with partial pivoting, the nth step in the reduction becomes

$$
\left.
\begin{aligned}
u_{in} &= a_{in} - \sum_{k=1}^{i-1} l_{ik} u_{kn} & i &= 1, n-1 \\[2ex]
a_{in} &= a_{in} - \sum_{k=1}^{n-1} l_{ik} u_{kn} & i &= n, N \\[2ex]
u_{nn} &= \underset{i}{MAX}\left| a_{in} \right| & i &= n, N \\[2ex]
l_{in} &= a_{in}/u_{nn} & i &= n+1, N
\end{aligned}
\right\} \quad n = 1, N
\qquad (2.30)
$$

where the search for the pivot element among the unreduced rows of column n (step 3 above) is culminated by one interchange of the nth row with the pivot row to produce u_{nn}. Note that since a single column at a time is processed, the entire column can be transferred to a double or extended precision vector for reduction and the result transferred back to the original array storage locations. This procedure in effect results in double or extended precision computation at the expense of only single column of double precision storage.

The variations of LU decomposition described thus far have reduced A exclusively by rows or by columns. There is no restriction preventing a combination of the two forms from being used. Such a choice might be dictated by the fact that compact storage assignments used in handling sparse matrices store elements above the diagonal of A by rows and below the diagonal of A by columns. The elements of U and L are then most easily computed in the order

$$u_{11}\ u_{12}\ u_{13}\ u_{14}\ \cdots\ u_{1N}$$

$$l_{21}\ l_{31}\ l_{41}\ l_{51}\ \cdots\ l_{N1}$$

$$u_{22}\ u_{23}\ u_{24}\ \cdots\ u_{2N}$$

$$l_{32}\ l_{42}\ l_{52}\ \cdots\ l_{N2}$$

$$u_{33}\ u_{34}\ \cdots\ u_{3N}$$

$$l_{43}\ l_{53}\ \cdots$$

$$\vdots$$

This variation is described at the n th step of the reduction by

$$\left. \begin{aligned} u_{nj} &= a_{nj} - \sum_{k=1}^{n-1} l_{nk} u_{kj} \qquad j = n, N \\[2em] l_{in} &= (a_{in} - \sum_{k=1}^{n-1} l_{ik} u_{kn})/u_{nn} \quad i = n+1, N \end{aligned} \right\} \quad n = 1, N \qquad (2.31)$$

2.4 Determinants

It will be shown in a later chapter that a very effective method for computing poles and zeros of small-signal transfer functions requires the computation of determinants. With the single exception of Crout's method, all of the above procedures facilitate this computation, for each procedure can be shown to be the equivalent of applying a series of linear transformations to the original system of equations (2.1) in order to obtain an equivalent triangularized system of equations (2.20). It can be proven that subject to such linear transformations,

$$|A| = |U|$$
(2.32)

But, it can easily be seen that $|U|$ is given simply by the product of terms on the diagonal since U is upper-triangular. Thus

$$|A| = |U| = \prod_{i=1}^{N} u_{ii}$$
(2.33)

Clearly, Crout's method in which $u_{ii} = 1$, $i = 1,N$, cannot give the correct value for the determinant. The problem is that the required division step in the calculation of u_{ij} represents a violation of the rules for linear transformations.

2.5 Accuracy Enhancement

The use of pivoting as a means of improving solution accuracy in LU decomposition has been mentioned previously. The problem of computing with numbers of finite precision can be illustrated via the following example taken from Freret [E 12] where it is assumed three-decimal floating arithmetic is used

$$\begin{bmatrix} .000100 & 1.00 \\ 1.00 & 1.00 \end{bmatrix} \begin{bmatrix} X_1 \\ X_2 \end{bmatrix} = \begin{bmatrix} 1.00 \\ 2.00 \end{bmatrix}$$
(2.34)

After one step of Gaussian elimination, the system becomes

$$\begin{bmatrix} .000100 & 1.00 \\ & -10,000 \end{bmatrix} \begin{bmatrix} X_1 \\ X_2 \end{bmatrix} = \begin{bmatrix} 1.00 \\ -10,000 \end{bmatrix}$$
(2.35)

which yields the solution

$X_2 = 1.00$
$X_1 = 0.00$

an obviously incorrect result. On the other hand, if the rows are interchanged such that in column 1 the element of largest magnitude is placed on the diagonal, after one

Gaussian elimination step the system becomes

$$\begin{bmatrix} 1.00 & 1.00 \\ & 1.00 \end{bmatrix} \begin{bmatrix} X_1 \\ X_2 \end{bmatrix} = \begin{bmatrix} 2.00 \\ 1.00 \end{bmatrix} \qquad (2.37)$$

which yields the correct solution

$$X_2 = 1.00$$
$$X_1 = 1.00$$

Pivoting for this example has in effect helped control the range of the magnitudes of the terms in the equations which thereby results in improved accuracy.

The limitation of pivot strategies to row interchanges is usually referred to as partial pivoting. It preserves column order and hence the order of the unknowns in the solution vector. Complete pivoting involves finding the largest element in the as yet unreduced $(N-n) x (N-n)$ submatrix at the nth step, and the exchange of rows and columns such that it appears on the diagonal in the nth row and column. Because of the added complexity, complete pivoting is seldom used.

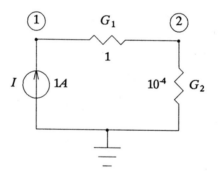

Figure 2.3 - Simple three element circuit

In general, pivoting or positioning for size is not amenable to processing with sparse matrices. The reason is that sparse matrix processing usually involves a complete determination of processing order and thus a determination of the location of all nonzero terms before the actual magnitudes of any terms are known. In this respect nodal analysis and modified nodal analysis possess the desirable property that the nodal admittance (conductance) matrix and the reduced matrix are usually diagonally dominant. This diagonal dominance results from the fact that for all basic elements under consideration, they appear on the diagonal with a plus (+) sign. Nonetheless, problems can still be encountered as illustrated via the following example again taken from Freret [E 13] The nodal equations for the simple three element circuit of Figure 2.3 can be written symbolically as

$$\begin{bmatrix} G_1 & -G_1 \\ -G_1 & G_1 + G_2 \end{bmatrix} \begin{bmatrix} V_1 \\ V_2 \end{bmatrix} = \begin{bmatrix} I \\ 0 \end{bmatrix}$$ (2.39)

If it is assumed that a computer implementation of a simulation program retains 4 significant figures, the equation (2.39) becomes

$$\begin{bmatrix} 1.000 & -1.000 \\ -1.000 & 1.000 \end{bmatrix} \begin{bmatrix} V_1 \\ V_2 \end{bmatrix} = \begin{bmatrix} 1.000 \\ 0 \end{bmatrix}$$ (2.40)

which after one step of Gaussian elimination results in

$$\begin{bmatrix} 1.000 & -1.000 \\ 0 & 0 \end{bmatrix} \begin{bmatrix} V_1 \\ V_2 \end{bmatrix} = \begin{bmatrix} 1.000 \\ 1.000 \end{bmatrix}$$ (2.41)

which yields the incorrect solution.

$$V_2 = \infty$$
$$V_1 = \infty$$

For this sample, pivoting won't help because the significant piece of data reflecting the presence of G_2 was irretrievably lost in assembling (2.40). The result is a current source forced into an open circuit. It has recently been demonstrated that by exploiting a property of the indefinite admittance matrix as pointed out by Bingham, the above problem can be resolved.[E7] Recall that the indefinite admittance matrix is obtained by appending the row and column corresponding to the ground, datum or reference node to the (definite) admittance matrix. Symbolically

$$Y_a = \begin{bmatrix} y_{11} & y_{12} & \cdot & y_{1n} & | & y_{10} \\ y_{21} & & & \cdot & | & \cdot \\ \cdot & & & \cdot & | & \cdot \\ \cdot & & & \cdot & | & \cdot \\ y_{n1} & \cdot & \cdot & y_{nn} & | & \cdot \\ -- & -- & -- & -- & | & -- \\ y_{01} & \cdot & \cdot & \cdot & | & y_{00} \end{bmatrix}$$ (2.43)

The indefinite admittance matrix is characterized by the properties

1)   The sum of the elements in any row is zero.

2)   The sum of the elements in any column is zero.

Property 1 can be used to more accurately calculate y_{nn} as follows

$$y_{nn} = - \sum_{\substack{j=1 \\ j \neq n}}^{N+1} y_{nj} \tag{2.44}$$

Since property 1 is preserved through linear transformations,, U_{nn} can be calculated at each stage of an LU decomposition as

$$u_{nn} = - \sum_{j=n+1}^{N+1} u_{nj} \tag{2.45}$$

The previous example now proceeds as follows: The initial system is written as

$$\begin{bmatrix} G_1 & -G_1 & | & 0 \\ -G_1 & G_1+G_2 & | & -G_2 \\ -- & ------ & | & -- \\ 0 & -G_2 & | & G_2 \end{bmatrix} \begin{bmatrix} V_1 \\ V_2 \\ -- \\ V_0 \end{bmatrix} = \begin{bmatrix} I \\ 0 \\ -- \\ -I \end{bmatrix} \tag{2.46}$$

After one step of Gaussian elimination

$$\begin{bmatrix} G_1 & -G_1 & | & 0 \\ 0 & y_{22} & | & -G_2 \\ -- & -- & | & -- \end{bmatrix} \begin{bmatrix} V_1 \\ V_2 \\ -- \end{bmatrix} = \begin{bmatrix} I \\ I \\ -- \end{bmatrix} \tag{2.47}$$

where

$$y_{22} = - \sum_{j=3}^{3} y_{2j} = -(-G_2) = G_2 \tag{2.48}$$

Again with 4 significant figures used

$$\begin{bmatrix} 1.000 & -1.000 & | & 0 \\ -1.000 & 1.000 & | & -0.0001 \end{bmatrix} \begin{bmatrix} V_1 \\ V_2 \end{bmatrix} = \begin{bmatrix} 1.000 \\ 0 \end{bmatrix} \tag{2.49}$$

which after one step of elimination yields

$$\begin{bmatrix} 1.000 & -1.000 & | & 0 \\ 0 & 0.0001 & | & -0.0001 \end{bmatrix} \begin{bmatrix} V_1 \\ V_2 \end{bmatrix} = \begin{bmatrix} 1.000 \\ 1.000 \end{bmatrix} \tag{2.50}$$

and now results in the more correct solution

$$V_2 = 10,000 \qquad V_1 = 10,000 \tag{2.51}$$

While the above procedure was illustrated for the case of nodal analysis, it can be applied to any system of linear equations and was done so quite commonly for hand computations schemes as a check on arithmetic error. Nonetheless, the application to nodal analysis retains an unmatched advantage in that the $N+1$ column can be obtained directly from the circuit description by inspection without consideration of other terms in a row. Additionally terms almost always appear with a negative sign resulting in little chance of accuracy loss through computing a small difference of large numbers. Finally, note that the above method is incompatible with computation of elements of U by columns.

2.6 Iterative Methods

While not used under normal circumstances for solving linear circuit equations, iterative methods have typically been used to improve or refine solutions first obtained by direct solution via Gaussian elimination or LU transformation. Their accuracy derives from the fact that there is virtually no propagation of error such as may occur in matrix triangularization.

As with the LU transformation approach, iterative methods are also based upon a factorization of A into two constituent parts. However, the factorization is of the form

$$A = P - Q \tag{2.52}$$

The system is then written as

$$P X = Q X + B \tag{2.53}$$

A sequence of iterates $\cdots X^{(\nu)}, X^{(\nu+1)}, \cdots$ is derived from an initial guess $X^{(0)}$ by the recursion

$$P X^{(\nu+1)} = Q X^{(\nu)} + B \tag{2.54}$$

which clearly requires

$$\det |P| \neq 0 \tag{2.55}$$

The first method to be considered, attributed to Jacobi, is based on simultaneous iteration where

$$P = \begin{bmatrix} a_{11} & & & & \\ & \cdot & & 0 & \\ & & \cdot & & \\ & & a_{nn} & & \\ & & & \cdot & \\ & 0 & & \cdot & \\ & & & & a_{NN} \end{bmatrix} \qquad (2.56a)$$

$$Q = P - A = \begin{bmatrix} 0 & a_{12} & & & \\ a_{21} & 0 & & & \\ & & \cdot & & \\ & & & \cdot & \\ & & & & \cdot \\ a_{N1} & \cdot & \cdot & & 0 \end{bmatrix} \qquad (2.56b)$$

It can be seen from (2.52) and (2.54) that the component $X_n^{(\nu+1)}$ at the $\nu+1$st iteration is given by

$$X_n^{(\nu+1)} = \frac{1}{a_{nn}} \left[b_n - \sum_{\substack{j=1 \\ j \neq n}}^{N} a_{nj} X_j^{(\nu)} \right] \qquad (2.57)$$

The operational count for the Jacobi iteration is N^2 operations per iterations.

The second and final method to be considered is that of successive or sequential iteration. This Gauss-Seidel method differs from Jacobi iteration in the following way. It is clear in the Jacobi iterations that some components of $X^{(\nu+1)}$ are known but not used while computing the remaining components. The Gauss-Seidel method takes into account the new information contained in these components and results in an iterative formula as follows

$$X_n^{(\nu+1)} = \frac{1}{a_{nn}} \left[b_n - \sum_{j=1}^{n-1} a_{ij} X_j^{(\nu+1)} - \sum_{j=n+1}^{N} a_{ij} X^{(\nu)} \right] \qquad (2.58)$$

Here, the factoring of A is given by

$$P = \begin{bmatrix} a_{11} & & & \\ a_{21} & a_{22} & & \\ \cdot & & \cdot & \\ \cdot & & & \cdot \\ \cdot & & & \cdot \\ a_{N1} & \cdot & \cdot \cdot & a_{NN} \end{bmatrix} \qquad Q = A - P \qquad (2.59)$$

The conditions assuring convergence of either of these methods have not been considered here. Basically, a strong or dominant diagonal is required. Under convergent conditions, Gauss-Seidel iteration does converge faster than Jacobi iteration. The basic problem is that for nonlinear circuits undergoing switching transients, the diagonal is not always dominant.

3. SPARSE MATRIX METHODS

In view of the previously derived N^3 dependence of the long operations count on the number of equations to be solved via Gaussian elimination, computation time can be expected to increase significantly as larger circuits with more nodes are considered. As a consequence, much attention has been focused on taking advantage of sparsity in nodal admittance matrix and other formulations. There are three associated savings. First, efficient means have been found by which only the non-zero entries of the matrix need be stored, thus effecting a savings in memory requirements. Second, it is possible to process only the non-zero entries at each step in the triangular reduction and back substitution. Finally, the order in which variables are eliminated can be chosen to preserve sparsity. The long operations count and computation time are then reduced. This savings becomes even more significant when the same equations must be solved many times as in transient or multi-frequency analyses. The optimal order for eliminating variables need only be determined once.

3.1 Sparse Matrix Storage

Two different techniques which have both been used for storage and handling of sparse matrices are presented in this section. The first referred to for want of a better name as "row-column indexing" has been used in CANCER [C7] and SPICE1 [C8]. The second, a "threaded list" technique has been used in SLIC [C5] and NICAP [C4].

The description and example of row column indexing which follows is taken from Nagel [C11]. The approach is illustrated in Figure 3.1. In general, integer pointers are used to access uniquely defined memory locations for each non-zero element in the coefficient matrix. Diagonal terms, usually non-zero and hence non-sparse, need no reference pointer and are assigned locations in the array AD. Similarly, the excitation vector also is usually non-zero due to the presence of Norton equivalent sources and thus is also stored as the nonsparse vector B. The remaining non-zero off-diagonal elements are stored in vector form by upper-triangular row and then lower-triangular column. The IUR array is used to access upper-triangular row entries. The number of entries in row i is given by $IUR(i+1) - IUR(i)$. Thus for example, there are zero entires to the right of the diagonal in row 1, one entry in row 2, two entries in row 3 and zero entries in rows 4 and 5. The column number for each row entry is stored in the IO array and its value in the AO array. Thus for row 2,

$$\begin{bmatrix} 11 & 0 & 0 & 0 & 0 \\ 0 & 8 & 0 & 0 & -6 \\ 0 & -2 & 6 & -1 & -3 \\ -4 & 0 & 0 & 5 & 0 \\ 0 & 0 & -5 & 0 & 7 \end{bmatrix} \begin{bmatrix} X_1 \\ X_2 \\ X_3 \\ X_4 \\ X_5 \end{bmatrix} = \begin{bmatrix} 1 \\ 3 \\ 5 \\ 2 \\ 4 \end{bmatrix}$$

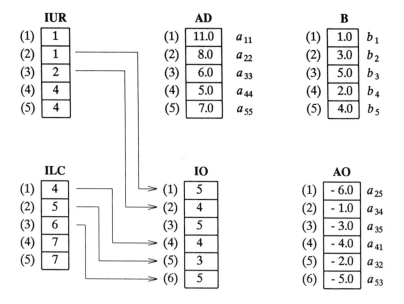

Figure 3.1 - Sparse matrix storage via row-column indexing

from $IUR(2) = 1$, it can be seen that $IO(1) = 5$ indicates the 5th column such that $a_{25} = -6.0$. Similarly from $IUR(3) = 2$, it can be seen that off-diagonal row terms for row 3 are stored beginning at $IO(2) = 4$ and extend through $IO(3) = 5$, since $(IUR(4) - IUR(3) = 2)$, thus giving $a_{34} = -1.0$ and $a_{35} = -3.0$. In a similar fashion, ILC points to the beginning of the lower-triangular column entries below the diagonal. For those familiar with assembly language programming on minicomputers, it can be seen that this system is similar to indirect addressing in that for row i or column j, the location in IO and AO of the first non-zero entry is given by $IUR(i)$ and $ILC(j)$.

Prior to a discussion of the merits of row-column indexing, it is useful to introduce the threaded list technique as well so that the two may be compared. A threaded list system is illustrated in Figure 3.2 for the same example as before. Again unique storage locations are provided for each non-zero term in the array $VALU$. The arrays $IROW$ and $JCOL$ record the i,j coordinates of each non-zero term in the system of equations. Finally, the arrays IPT and JPT point to the row i, of the next element in a column and the column j, of the next element in a row respectively. Values of IPT and JPT equal to zero indicate the end of a column or row respectively. Thus to scan down column 1, $IPT(1) = 6$ indicates the first non-zero entry is at location 6 where $IROW(6) = 1$ and $JCOL(6) = 1$ indicate a_{11}. $IPT(6) = 7$ points to the next entry at location 7 where $IROW(7) = 4$ and $JCOL(7) = 1$ indicate a_{41}. $IPT(7) = 0$ indicates that this was the last non-zero term in the column. Similarly to scan across row 3, one would begin at $JPT(3) = 9$ where $IROW(9) = 3$ and $JCOL(9) = 2$ indicate a_{32}, etc.

Note that as illustrated in Figure 3.2 and described above, this particular threaded list scheme is bi-directional. That is, from any element, one can proceed to the element below it via IPT and to the element to the right of it via JPT. This feature greatly facilitates the implementation of Gaussian elimination for equation solution.

The individual arrays or vectors $IROW$ and $JCOL$ for storing row and column locations respectively can be combined into a single array $IJSUM = IROW + JCOL$ to conserve storage. In scanning down a column, the column number is then subtracted from $IJSUM$ to determine each unique row number. Similarly, to scan across a row, the row number is subtracted from $IJSUM$ to determine each unique column number. Thus to proceed cross row 3, one would begin at $JPT(3) = 9$ where $IJSUM(9) - 3 = 5 - 3 = 2$ indicates a_{32}, etc. as before.

Note that, if desired, a similar technique could be used with IPT to allow moving up a column as well as down by storing the sum of the location of the previous entry and the location of the next entry. To move down, subtract the location of the previous entry to get the next entry. To move up, the procedure is reversed. A similar technique can be used with JPT to allow linking from right to left as well as left to right. Thus with no increase in storage, linking of any term in the list to any of its four nearest neighbors can be accomplished.

$$\begin{bmatrix} 11 & 0 & 0 & 0 & 0 \\ 0 & 8 & 0 & 0 & -6 \\ 0 & -2 & 6 & -1 & -3 \\ -4 & 0 & 0 & 5 & 0 \\ 0 & 0 & -5 & 0 & 7 \end{bmatrix} \begin{bmatrix} X_1 \\ X_2 \\ X_3 \\ X_4 \\ X_5 \end{bmatrix} = \begin{bmatrix} 1 \\ 3 \\ 5 \\ 2 \\ 4 \end{bmatrix}$$

	IPT	IROW	JCOL	JPT	VALU		IJSUM
(1)	6	-	-	6	1.0	b_1	-
(2)	8	-	-	8	3.0	b_2	-
(3)	11	-	-	9	5.0	b_2	-
(4)	13	-	-	7	2.0	b_2	-
(5)	10	-	-	12	4.0	b_2	-
(6)	7	1	1	0	11.0	a_{11}	2
(7)	0	4	1	15	-4.0	a_{41}	5
(8)	9	2	2	10	8.0	a_{22}	4
(9)	0	3	2	11	-2.0	a_{32}	5
(10)	14	2	5	0	-6.0	a_{25}	7
(11)	12	3	3	13	6.0	a_{33}	6
(12)	0	5	3	16	-5.0	a_{53}	8
(13)	15	3	4	14	-1.0	a_{34}	7
(14)	16	3	5	0	-3.0	a_{35}	8
(15)	0	4	4	0	5.0	a_{44}	8
(16)	0	5	5	0	7.0	a_{55}	10

Figure 3.2 - Sparse matrix storage via bi-directional threaded list

There are several considerations to be weighed in choosing one storage technique over the other. Among them are storage requirements, processing efficiency and flexibility. In terms of storage, the row-column indexing does usually require less memory for pointer storage -- $N+P$ for general matrices and $(N+P)/2$ for matrices symmetric about the diagonal where N is the number of equations and P is the number of non-zero terms. The bi-directional threaded list scheme requires $2N+3P$ in either case while a uni-directional threaded list requires $N+2P$ locations for pointer storage. Historically, on byte oriented machines, pointers were usually stored in two byte (16 bit) words, while on word oriented machines they were packed two per word (32 bit), three per word (36 bit) or five per word (60 bit).

In terms of flexibility, the bi-directional threaded-list usually provides more, particularly if several basic subroutines for list manipulation are developed -- FILLIJ(I,J) to record a non-zero entry a_{ij}, LOCIJ(I,J) to test for the presence of a non-zero entry at a_{ij}, DELETEIJ(I,J) to delete row I and column J from the list, ADDIJ(I1,J2,I2,J2) to add row I1 to row I2 and column J1 to column J2, SWAPIJ(I1,J1,I2,J2) to exchange row I1 with row I2 and column J1 with column J2 and finally LISTIJ to list the entire threaded list for debugging purposes.

A particularly nice feature of the threaded list approach is the ease with which new non-zero terms may be added to the matrix. For instance, to modify the example of Figure 3.2 in order to add a non-zero term a_{23}, the following changes and additions must be made: $IPT(3) = 17$, $IPT(17) = 11$, $IROW(17) = 2$, $JCOL(17) = 3$, (or $IJSUM(17) = 5$), $JPT(8) = 17$, $JPT(17) = 10$. For the same addition to the row-column indexing approach of the example in Figure 3.1, $IUR(3) - IUR(5)$ must incremented by 1, all of ILC must be incremented by 1 and finally the entire contents of $I0$ and $A0$ must be shifted downward one location to make room for a_{23} ahead of a_{25}.

Recall that the basic algorithm for performing an LU-factorization can be described in equation form as

$$
\left.
\begin{aligned}
l_{in} &= a_{in}^{(n-1)}/a_{nn}^{(n-1)} && i = n+1,N \\[2ex]
u_{nj} &= a_{nj}^{(n-1)} && j = n,N \\[2ex]
a_{ij}^{(n)} &= a_{ij}^{(n-1)} - l_{in}\,u_{nj} && \begin{aligned} i &= n+1,N \\ j &= n,N \end{aligned}
\end{aligned}
\right\} \quad n = 1,N \qquad (3.1)
$$

such that terms are calculated in the order

$$l_{21} \quad l_{31} \quad l_{41} \quad \cdots \quad l_{N1}$$

$$u_{11} \quad u_{12} \quad u_{13} \quad \cdots \quad u_{1N}$$

$$a_{22}^{(1)} \quad a_{23}^{(1)} \quad a_{24}^{(1)} \quad \cdots \quad a_{2N}^{(1)}$$

$$\cdot$$
$$\cdot$$
$$\cdot$$

$$a_{N2}^{(1)} \quad a_{N3}^{(1)} \quad a_{N4}^{(1)} \quad \cdots \quad a_{NN}^{(1)}$$

$$l_{32} \quad l_{42} \quad \cdots \quad l_{N2}$$

$$u_{22} \quad u_{32} \quad \cdots \quad u_{2N}$$

$$a_{33}^{(2)} \quad a_{34}^{(2)} \quad \cdots \quad a_{3N}^{(2)}$$

$$\cdot$$
$$\cdot$$
$$\cdot$$

It can be seen that the row-column indexing approach of storing upper-triangular non-zero terms consecutively by rows and lower-triangular terms consecutively by columns greatly facilitates the calculation of the l and u terms. The problem calculation within the procedure is that of computing updated values of $a_{ij}^{(n)}$. For sparse systems, these terms are located virtually at random within the AO array. They can only be located through the existing pointer arrays by starting from $IUR\,(i)$ in IO and looking for j in IO if $i \geq j$ or starting from $ILC\,(j)$ in IO and looking for i if $i \geq j$. This procedure may require sufficient effort to warrant alternative methods.

One such technique is to record, for each step in the reduction, the location in AO (and IO) of $a_{ij}^{(n-1)}$, the term to be updated, in a MEMO array. That is, in performing the operation

$$a_{ij}^{(n)} = a_{ij}^{(n-1)} - l_{in}\, u_{nj} \qquad \begin{array}{l} j = n\,,N \\ i = n+1,N \end{array} \tag{3.2}$$

the terms l_{in} and u_{nj} are obtained sequentially from AO because of the order of storage, while the location of a_{ij} in AO is obtained from MEMO. This procedure implies that the size of MEMO be equal to the number of operations required for the reduction which, it will be pointed out later, is of the order of N^2. Nagel was found that the use of such a MEMO array can result in a 10% to 15% savings in execution speed.

It was pointed out that Gaussian elimination or LU factorization can be performed in many different sequences of steps. A natural question to ask is whether any particular variation is characterized by a sequence of steps which lends itself to a better sparse-matrix implementation than the basic LU reduction of (3.1). The answer is yes. Note that the reason for considering a MEMO array was execution of the operation of (3.2). The method (2.31) described in equation form by

$$\left.\begin{array}{ll} u_{nj} = a_{nj} - \displaystyle\sum_{k=1}^{n-1} l_{nk}\, u_{kj} & j = n\,,N \\[4mm] l_{in} = a_{in} - \displaystyle\sum_{k=1}^{n-1} l_{ik}\, u_{kn}/u_{nn} & i = n+1,N \end{array}\right\} \quad n = 1,N \tag{3.3}$$

generates terms in the sequence

$$u_{11}\ u_{12}\ u_{13}\ u_{14}\ \cdots\ u_{1N}$$

$$l_{21}\ l_{31}\ l_{41}\ l_{51}\ \cdots\ l_{N1}$$

$$u_{22}\ u_{23}\ u_{24}\ \cdots\ u_{2N}$$

$$l_{32}\ l_{42}\ l_{52}\ \cdots\ l_{N2}$$

$$u_{33}\ u_{34}\ \cdots\ u_{3N}$$

$$l_{43}\ l_{53}\ \cdots$$

$$\vdots$$

The calculations required for computation of these terms are illustrated in Figure 3.3. Here again, however, if one examines closely the calculations for $n = 3$, it can be seen that in the calculation for u_{33}, it is impossible for the row-column indexing to proceed

$$u_{11} = a_{11}$$
$$u_{12} = a_{12}$$
$$\cdot$$
$$\cdot$$
$$l_{21} = a_{21}/u_{11}$$
$$l_{31} = a_{31}/u_{11}$$
$$\cdot$$
$$\cdot$$

$\left. \phantom{\begin{array}{c}a\\a\\a\\a\\a\\a\\a\\a\end{array}} \right\} \quad n = 1$

$$u_{22} = a_{22} - l_{21}u_{12}$$
$$u_{23} = a_{23} - l_{21}u_{13}$$
$$\cdot$$
$$\cdot$$
$$l_{32} = (a_{32} - l_{31}u_{12})/u_{22}$$
$$l_{42} = (a_{42} - l_{41}u_{12})/u_{22}$$
$$\cdot$$
$$\cdot$$

$\left. \phantom{\begin{array}{c}a\\a\\a\\a\\a\\a\\a\\a\end{array}} \right\} \quad n = 2$

$$u_{33} = a_{33} - l_{31}u_{13} - l_{32}u_{23}$$
$$u_{34} = a_{34} - l_{31}u_{14} - l_{32}u_{24}$$
$$\cdot$$
$$\cdot$$
$$l_{43} = (a_{43} - l_{41}u_{13} - l_{42}u_{23})/u_{33}$$
$$l_{53} = (a_{53} - l_{51}u_{13} - l_{52}u_{23})/u_{33}$$
$$\cdot$$
$$\cdot$$

$\left. \phantom{\begin{array}{c}a\\a\\a\\a\\a\\a\\a\\a\end{array}} \right\} \quad n = 3$

Figure 3.3 - Calculations required in a variation of LU-factorization

directly from u_{13} to u_{23} because u_{ij} is stored by rows, not columns.

On the other hand, the bi-directional threaded list approach does provide this ability. Consequently, each new operation in the threaded list approach requires only a single table lookup without the necessity of a MEMO array. In fact, the storage required for the bi-directional threaded list is usually less than that required for row-column indexing with a MEMO array.

The trade-off thus remains between storage versus speed. It can be pointed out that for nonlinear dc and/or transient analysis, the time spent solving linear equations varies between 10% and 20% of the time per Newton iteration. On the other hand, for small-signal ac and/or pole-zero analysis, the percent of time per frequency point or time per root iteration spent in linear equation solution dominates. In either of these cases, speed often outweighs storage and the use of a MEMO array or the threaded list approach does become desirable. Finally, it is the rule rather than the exception that for pole-zero analyses, a nonsymmetric non-zero structure is encountered. This fact negates the advantage of row-column indexing in handling symmetric matrix structures.

Two additional techniques regarding sparse matrix implementations of equation solution should also be mentioned. The first is the use of assembly language programming in the solution subroutine so as to optimize register usage and minimize redundant loads and stores. The second techniques is the generation of explicit machine code which contains no loops or checking for number of terms, locations and ends of rows and columns. The former has been found by Nagel to result in a savings of approximately 50% in execution speed at no increase in storage, while the latter was found to result in a savings of approximately 70% in execution speed at a sometimes tremendous increase in storage requirements.

3.2 Optimal Ordering

As initially pointed out, a third savings accrued from sparse matrix techniques involves the selection of pivot order to preserve sparsity. The desirability of establishing an optimal pivot order can be seen from the two examples of Figure 3.4. The ordering of Figure 3.4(a) is the worst possible. All previously zero terms become non-zero during the course of a Gaussian elimination. The best possible ordering for this example is shown in Figure 3.4(b). This ordering preserves the sparsity of the equations. All terms originally zero remain zero.

$$
\begin{matrix}
a: \\
b: \\
c: \\
d:
\end{matrix}
\begin{bmatrix}
x & x & x & x \\
x & x & 0 & 0 \\
x & 0 & x & 0 \\
x & 0 & 0 & x
\end{bmatrix}
\begin{bmatrix}
V_a \\
V_b \\
V_c \\
V_d
\end{bmatrix}
=
\begin{bmatrix}
x \\
x \\
x \\
x
\end{bmatrix}
\quad => \quad
\begin{bmatrix}
x & x & x & x \\
x & x & x & x \\
x & x & x & x \\
x & x & x & x
\end{bmatrix}
\begin{bmatrix}
V_a \\
V_b \\
V_c \\
V_d
\end{bmatrix}
=
\begin{bmatrix}
x \\
x \\
x \\
x
\end{bmatrix}
$$

Before   After

(a)

$$
\begin{matrix}
a: \\
b: \\
c: \\
d:
\end{matrix}
\begin{bmatrix}
x & 0 & 0 & x \\
0 & x & 0 & x \\
0 & 0 & x & x \\
x & x & x & x
\end{bmatrix}
\begin{bmatrix}
V_a \\
V_b \\
V_c \\
V_d
\end{bmatrix}
=
\begin{bmatrix}
x \\
x \\
x \\
x
\end{bmatrix}
\quad => \quad
\begin{bmatrix}
x & 0 & 0 & x \\
0 & x & 0 & x \\
0 & 0 & x & x \\
x & x & x & x
\end{bmatrix}
\begin{bmatrix}
V_a \\
V_b \\
V_c \\
V_d
\end{bmatrix}
=
\begin{bmatrix}
x \\
x \\
x \\
x
\end{bmatrix}
$$

Before   After

(b)

Figure 3.4 - (a) Worst possible ordering for Gaussian Elimination versus (b) best possible ordering

    Numerous ordering algorithms have been proposed. Four of these which have been proposed for use with diagonal pivoting are summarized in Table 3.1. [G1,G6,G13,G15] Here, $NZUR$ represents the number of non-zero terms (including the diagonal) in an upper-triangular row, $NZLC$ represents the number of non-zero terms (including the diagonal) in a lower-triangular column. Thus, in the Markowitz algorithm, the next pivot element is chosen such that the product of its off diagonal terms $(NZUR-1)*(NZLC-1)$, is minimized. In case of ties, the pivot element for which $NZLC$ is minimized is chosen. Nonzero terms in the rows and columns of previously selected pivot elements are ignored. This algorithm is based upon the fact that the maximum number of fill-ins which could possibly occur for a particular pivot selection is $(NZUR-1)*(NZLC-1)$. Note that after each pivot element is selected, the sparse matrix pointer system must be updated to reflect any generated fill-ins. The same requirement is true for the remaining algorithms.

MARKOWITZ:	1)	$\min[(NZUR-1)*(NZLC-1)]$
	2)	$\min(NZLC)$
BERRY:	1)	$\min(NFILL)$
	2)	$\max(NZUR)$
	3)	$\max(NZLC)$
HSIEH:	1)	Row Singletons $(NZUR = 1)$
	2)	Column Singletons $(NZLC = 1)$
	3)	$(NZUR,NZLC) = (2,2),(2,3),(2,4),(4,2),(3,3)$
	4)	$\min[(NZUR-1)*(NZLC-1)]$
	5)	$\max(NZLC)$
NAHKLA:	1)	$\min(NFILL)$
	2)	$\min(NZUR)$
	3)	$\min(NZLC)$
where		$NZUR$ = Nonzero terms in upper triangular row
		$NZLC$ = Nonzero terms in lower triangular column
		$NFILL$ = Number of actual fill-ins at each step

TABLE 3.1 - Comparison Summary of Four Ordering Algorithms for Diagonal Pivoting

The scheme proposed by Berry carries the above procedure a step further by computing for each potential pivot the actual number of fill-ins, $NFILL$, which would result. In case of ties, the pivot element for which $NZUR$ is maximized and then $NZLC$ is maximized is chosen. These latter conditions are based on the assumption that a removal from future consideration of as many non-zero terms as possible will result in fewer total fill-ins. Hsieh has pointed out that where the Markowitz scheme is based on a probabilistic approach, the Berry scheme is basically deterministic. The price paid for such added sophistication can be considerable. For example, it has been pointed out that for nonsparse systems, the number of operations required is of the order of $N^3/3$. Since sparse matrix ordering requires a pseudo-Gaussian elimination, it can be expected that $O(N^3)$ searches will be required to examine the number of possible fill-ins in selecting the first pivot. Similarly, for the second pivot, $O((N-1)^3)$ searches will be required, etc. As a consequence, the total number of searches will be $O(N^3!)$, a very large number for even moderate values of N.

To illustrate the above observation, ordering times and number of fill-ins for a circuit containing 69 transistors and 160 equations were compared using the SLIC program. The results are shown in Table 3.2. (Note that a slow computer was used to begin with!) It can be seen that the Markowitz algorithm performs as well as Berry's algorithm and is much faster. Further comparisons are cited below.

	Markowitz	Berry
Transistors	69	69
Equations	160	160
Fill-ins	314	310
Non-zero Terms	1130	1124
Operations	2370	2346
Ordering Time	51 sec	780 sec

TABLE 3.2 - Comparison of Efficiency of Berry and Markowitz Ordering Algorithms

Hsieh has carried the idea of ordering in a probabilistic sense even further. His results are summarized in the third algorithm of Table 3.1. Here row singletons and column singletons (no possible fill-ins) are pro accessed first. Next, pivots are selected such that the number of non-zero row and column terms agree with the prescribed order of step 3. When no more such combinations exist, a Markowitz criteria is applied and finally, pivot elements for which $NZLC$ is maximized are chosen. Prior to citing Hsieh's comparative results, it is interesting to examine the order of selection for different combination of $NZUR$ and $NZLC$ between Hsieh and Markowitz as illustrated in Figure 3.5. Careful inspection reveals that the only difference outside of the circled terms is that one is a mirror image about the diagonal of the other. This difference results primarily from the choice of secondary selection criteria. As a consequence it is to be expected that the difference between the two schemes is negligible. (On a more practical note, it can be seen that merely by replacing $NZLC$ with $NZUR$ in the secondary Markowitz criteria, one obtains the Hsieh ordering.) Hsieh's comparative results are summarized in Table 3.3. Little difference is seen between his and the Markowitz algorithm.

The fourth ordering algorithm in Table 3.1 is due to Nahkla. To first order it is the same as Berry's. As Hsieh and Markowitz differ by opposite choice of secondary criteria, so also do Nahkla and Berry. Again the end result is a negligible difference. This fact is brought out by the comparative study done by Nagel and summarized in Table 3.4. For this study, samples of 20 matrices whose non-zero structure was generated randomly were used. Matrices were assumed symmetric about the diagonal. Nahkla's algorithm yielded only marginally better results yet required appreciably more time. A more extensive analysis involving up to 1000 equations led to similar conclusions.

Hsieh Ordering							
	NZLC						
	1	2	3	4	5	6	7
1	1	1	1	1	1	1	1
N 2	2	3	4	6	(9)	11	13
Z 3	2	5	(8)	14			
U 4	2	7	15	.			
R 5	2	10			.		
6	2	12				.	
7	2	16					.

(a)

Markowitz Ordering							
	NZLC						
	1	2	3	4	5	6	7
1	1	2	2	2	2	2	2
N 2	1	3	5	7	10	12	16
Z 3	1	4	(9)	15			
U 4	1	6	14	.			
R 5	1	(8)			.		
6	1	11				.	
7	1	13					.

(b)

Figure 3.5 - Comparison of (a) Hsieh ordering and (b) Markowitz ordering for combinations of NZLC and NZUR

	Berry	Markowitz	Hsieh
Equations	20	20	20
Fill-ins	5	5	5
Non-zero Terms	50	50	50
Operations	17	17	17
Ordering Time	0.1 sec	0.01 sec	0.01 sec
Equations	172	172	172
Fill-ins	258	271	271
Non-zero Terms	782	795	795
Operations	938	980	979
Ordering Time	67.44 sec	1.38 sec	1.03 sec
Equations	--	813	813
Fill-ins	--	1371	1371
Non-zero Terms	--	3896	3896
Operations	--	4970	4965
Ordering Time	--	29.17 sec	27.61 sec

TABLE 3.3   - Hsieh's Comparison of the Berry, Markowitz and Hsieh Ordering Algorithms

		50 Equations	100 Equations
Terms:	Markowitz	318.3 ± 35.2	783.5 ± 72.6
	Berry	317.1 ± 34.1	779.1 ± 70.4
	Nahkla	315.8 ± 34.6	778.5 ± 70.9
Operations:	Markowitz	943.6 ± 192.5	3015.9 ± 582.7
	Berry	933.9 ± 187.0	2965.0 ± 598.9
	Nahkla	925.7 ± 189.2	2962.9 ± 556.1
Time:	Markowitz	0.109 ± 0.023	0.446 ± 0.095
	Berry	0.528 ± 0.139	3.189 ± 0.867
	Nahkla	0.548 ± 0.150	3.235 ± 0.877

TABLE 3.4   - Comparison of Markowitz, Berry and Nahkla Ordering Algorithms from Nagel

As previously mentioned the four algorithms just described presume diagonal pivoting. For nodal and modified nodal such an *a priori* approaches are often characterized by a strong or dominant diagonal such that pivoting for accuracy is usually not required. However, modified nodal usually does require some auxiliary ordering rules which must be applied prior to any of the above mentioned algorithms. The modified nodal ordering algorithm described by Ho [E 12] is summarized as follows:

1) Row and column singletons due to grounded voltage sources and current sources.

2) Row interchange of node equation and branch relation for ungrounded voltage sources and inductors.

3) Pivoting on the diagonal (Markowitz,etc.).

It is pointed out that since singletons appear as symmetrically located pairs about the diagonal, after all singletons are processed the original diagonal remains on the diagonal. Similarly the interchange of node equations with branch relations, while resulting in ±1's or the diagonal, leaves the remaining diagonal terms unaffected.

As might be expected, the sparse tableau formulation because of its added generality requires a more general ordering strategy analogous to complete pivoting. Nonzero terms are classified into variability types [E 9]; that is, the frequency with which they must be modified. For the purpose at hand, four variability types can be identified:

1) Topological--either +1 or -1.

2) Constant--independent of time or unknown.

3) Time Dependent--dependent upon time.

4) Nonlinear--dependent upon unknown vector.

In determining pivot order, weights are assigned to each of the four variability types, such that a pivot selection creating a new topological term will be given preference over a pivot selection creating a new nonlinear term which must be processed at every Newton iteration.

4. NONLINEAR EQUATION SOLUTION

The problem of finding a solution to nonlinear algebraic circuit equations arises in the simulation of large-signal dc circuit behavior, large-signal dc transfer characteristics and large-signal time domain (or transient) responses. Nonlinear solution algorithms attempt to converge toward a solution to the system of nonlinear equations by iteratively developing and solving systems of linear algebraic equations. The method by which these linearized equations are developed is determined by the nonlinear solution method used. In general, all such methods are referred to as functional iteration.

Given a system of nonlinear equations such as

$$g(V) = I \qquad (4.1)$$

which might be obtained from a nodal formulation, they can be expressed in the form

$$\overline{g}(V) = g(V) - I = 0 \qquad (4.2)$$

Thus, deriving a solution to the system of nonlinear equations (4.1) is equivalent to the general problem of finding the roots or zeros of $\overline{g}(V)$ in (4.2). The solution technique is to start from some initial set of values $V^{(0)}$ and to generate a sequence of iterates $\cdots, V^{(n-1)}, V^{(n)}, V^{(n+1)}, \cdots$ which converge to the solution $\overline{V}$. Functional iteration is characterized by a transformation of the system of equations (4.2) into an equivalent system given by

$$V = f(V) \qquad (4.3)$$

which has the same roots. Iterates are then generated according to

$$V^{(n+1)} = f(V^{(n)}) \qquad (4.4)$$

In this chapter, a number of methods for transforming (4.2) into the form of (4.3) are examined. These methods include the Newton-Raphson, the Secant and the Chord methods.

4.1 Newton-Raphson Iteration

Newton-Raphson iteration is most easily introduced by considering first the case of a single nonlinear equation of the form (4.2). The function $\overline{g}(V)$ can be expanded about some point V_0 in a Taylor series to obtain

$$\overline{g}(V) = \overline{g}(V_0) + (V - V_0)\overline{g}'(V_0) + \cdots = 0 \tag{4.5}$$

where the prime denotes differentiation with respect to V. If only first order terms are retained, a rearrangement of (4.5) yields

$$V = V_0 - \overline{g}(V_0)/\overline{g}'(V_0) \tag{4.6}$$

Compared with (4.4), the form of (4.6) suggests that a sequence of iterates might be generated by the following:

$$V^{(n+1)} = V^{(n)} - \overline{g}(V^{(n)})/\overline{g}'(V^{(n)}) \tag{4.7}$$

Equation (4.7) is the Newton-Raphson iteration function for the scalar case. Note that at the solution $\overline{V}, \overline{g}(\overline{V}) = 0$ and $V^{(n+1)} = V^{(n)} = \overline{V}$ as would be expected. The geometrical interpretation of (4.7) is illustrated in Figure 4.1 for the simple case of a current source driving an ideal diode. The line tangent to the nonlinearity at the point $(V^{(n)}, \overline{g}(V^{(n)}))$ has the slope $\overline{g}'(V^{(n)})$. Its intercept with the voltage axis defines the next voltage iterate in the sequence as shown in the figure.

The generalization of the Newton-Raphson procedure to a system of N equations is given by

$$V^{(n+1)} = V^{(n)} - J^{-1}(V^{(n)})\overline{g}(V^{(n)}) \tag{4.8}$$

where the Jacobian $J(V)$ of the function $\overline{g}(V)$ is given by

$$J(V) = \begin{bmatrix} \dfrac{\partial g_1}{\partial V_1} & \dfrac{\partial g_1}{\partial V_2} & \cdot & \dfrac{\partial g_1}{\partial V_N} \\ \cdot & & & \cdot \\ \cdot & & & \cdot \\ \dfrac{\partial g_N}{\partial V_1} & \cdot & \cdot & \dfrac{\partial g_N}{\partial V_N} \end{bmatrix} \tag{4.9}$$

A physical interpretation of the elements of $J(V)$ is brought out below.

The direct application of (4.8) necessitates the computation of the NxN Jacobian matrix. As indicated previously the operation count for inverting a matrix and multiplying the result of a vector is $N^3 + N^2$. An alternative procedure for obtaining new iterates is to solve the linear system of equations

$$J(V^{(n)})(V^{(n)} - V^{(n+1)}) = \overline{g}(V^{(n)}) \tag{4.10}$$

A second alternative procedure often used with nodal analysis is to employ the system of equations

$$J(V^{(n)})V^{(n+1)} = J(V^{(n)})V^{(n)} - \overline{g}(V^{(n)}) \tag{4.11}$$

The right-hand side of (4.11) is found to have a particularly simple interpretation as will now be illustrated.

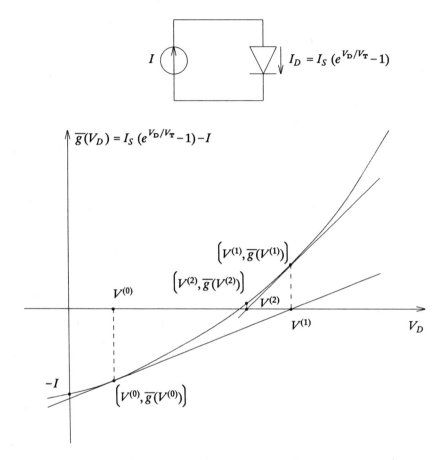

Figure 4.1 - Newton-Raphson iteration

A physical interpretation of the Jacobian matrix and the Newton-Raphson method can be made using the diode circuit of Figure 4.2(a). The exponential nonlinearity of the diode is linearized about some trial solution voltage V_0. This step is equivalent to expanding the nonlinearity in a Taylor series as indicated previously where only first-order terms are retained. The expansion is of the form

$$I_D = I_D \mid_{V=V_0} + (V-V_0)\frac{\partial I_D}{\partial V} \mid_{V=V_0} \tag{4.12}$$

$$= I_S(e^{V_0/V_T} - 1) + (V-V_0)\frac{I_S}{V_T}e^{V_0/V_T} \tag{4.13}$$

$$= I_{D0} + (V-V_0)G_{D0} \tag{4.14}$$

where I_{D0} is recognized as the current through the diode corresponding to the voltage V_0, and G_{D0} is recognized as the dynamic conductance corresponding to the voltage V_0. Since the diode characteristic as described by (4.14) has now been linearized, the diode of Figure 4.2(a) may be modeled in terms of a Norton-equivalent current source I_{DN0} in parallel with the conductance G_{D0}. As shown in Figure 4.2(b), I_{DN0} is given by

$$I_{DN0} = I_{D0} - G_{D0}V_0 \tag{4.15}$$

Hence, (4.14) may be written in terms of I_{DN0} as

$$I_D = G_{D0}V + I_{DN0} \tag{4.16}$$

The nodal equation for the complete linearized circuit of Figure 4.2(c) is

$$(G + G_{D0)}V = I - I_{DN0} \tag{4.17}$$

or in terms of iterate values is

$$(G + G_D{}^{(n)})V^{(n+1)} = I - I_{DN}^{(n)} \tag{4.18}$$

If (4.11) is compared with (4.18) the physical interpretations of the Jacobian and right-hand-side of (4.11) become more apparent. The Jacobian consists of the nodal conductance matrix of the linear elements of the circuit together with the linearized conductances associated with each nonlinear circuit element. The vector on the right-hand-side of (4.11) consists of independent source currents and the Norton-equivalent source currents associated with each nonlinear circuit element. Thus at each iteration in the Newton-Raphson procedure, the linearized conductances and Norton equivalent source currents must be recomputed and the linearized nodal conductance equations reassembled.

The Newton-Raphson procedure described above is illustrated in the flow-chart of Figure 4.3.

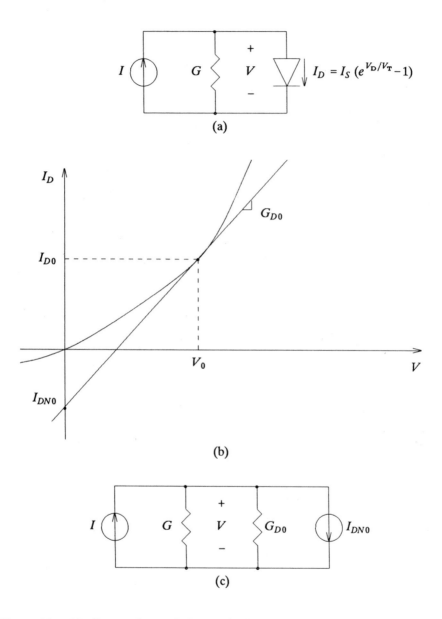

Figure 4.2 - Nonlinear dc analysis: (a) Diode circuit (b) Linearized diode approximation and (c) Linearized circuit model

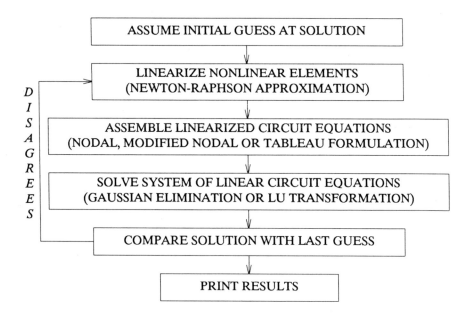

Figure 4.3 - Flow-chart for nonlinear analysis

4.2 Convergence and Termination

For the nonlinear dc analysis approach just outlined, the problem of convergence to a solution is now considered. Proofs that such an algorithm will converge depend upon *a priori* knowledge of an initial guess sufficiently close to the solution. Because this knowledge is usually not available in a nonlinear analysis, all techniques incorporated into analysis programs for improving convergence for the Newton-Raphson iteration technique are supported by purely empirical justification. The exponential nonlinearities usually associated with diodes and bipolar transistors are single-valued, monotonically increasing, continuous functions. However, these functions behave very violently. For large reverse bias, the slope approaches zero, while for large forward bias the exponential tends to infinity. The former case results in zero valued conductances and the latter in infinite conductances and Norton equivalent current sources. Convergence of the Newton-Raphson iterates may be slowed or the iteration procedure may be stopped when numbers overflow computer arithmetic capabilities. In what follows, several procedures for improving convergence properties as well as preventing overflows and underflows are presented.

Briefly stated, the problem, as illustrated in Figure 4.4, is to limit the positive excursion of junction voltage such that its exponentially dependent current does not

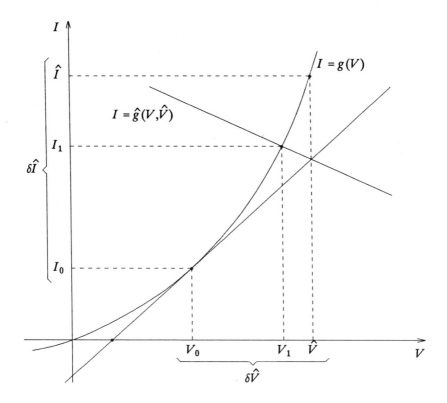

Figure 4.4 - Exponential junction voltage limiting

exceed a computer's range of floating-point number representations. In Figure 4.4 it is assumed that a diode exponential characteristic has been linearized at the point (V_0, I_0) and that a solution of the resulting linear circuit equation yields $\hat{V}$. It is desired to choose a new point for linearization

$$V_1 \leq \hat{V} \tag{4.19}$$

or

$$\delta V \leq \delta \hat{V} \tag{4.20}$$

such that

$$I_1 = I_S(e^{V_1/V_T} - 1) = g(V_1) \tag{4.21}$$

is representable within the range of valid floating point numbers. Numerous techniques for choosing V_1 have been proposed and/or tried. The simplest techniques include placing fixed bounds on the excursions δV or δI. A number of other techniques involve the creation of an auxiliary function

$$\hat{I} = \hat{g}(V, \hat{V}) \tag{4.22}$$

whose intersection with (4.21) defines the point (V_1, I_1). Several specific methods drawn from these two classes of techniques will be considered below. Two other techniques mentioned here but not considered further include: the selection of (V_1) such that the norm of $g(V)$ in (4.2) is forcibly reduced at each iteration or to ease into the solution by gradually turning on all sources starting from zero. The first technique is usually associated with Broyden [F4], Brown [F5], and Davidenko [F1] while the second technique, in effect a dc transfer characteristic, has been proposed by Cermak [F9] and others.

One of the most effective forms of the fixed bound approach was that implemented in the CANCER program by Nagel [C7]. It is summarized in Figure 4.5. In essence, for voltages less than $10V_T$, no limiting is used (i.e., $V_1 = \hat{V}$) while for voltages greater than $10V_T$, excursions are limited to $V_0 + 2V_T$. Special consideration is made for voltages at or near $10V_T$.

A second fixed-bound approach which has been tried in SLIC [C12] makes use of the hyperbolic tangent function and is summarized in Figure 4.6. It can be seen that since the hyperbolic tangent function ranges from -1 to +1, the maximum excursion of V_1 is $V_0 \pm 10V_T$. For small excursions $V_1 = \hat{V}$ since the slope of the hyperbolic tangent function for small arguments is one. Since a continuous function is used, no special cases need to be considered.

The simplest form of the auxiliary function technique, which was used in BIAS-3 [C3], has been referred to as an alternating basis technique in which current or voltage is used as a basis for iteration: iteration on current is used for increasing voltages and iteration on voltage is used for decreasing voltages. This method is summarized in Figure 4.7. For this method, the auxiliary function defined to be

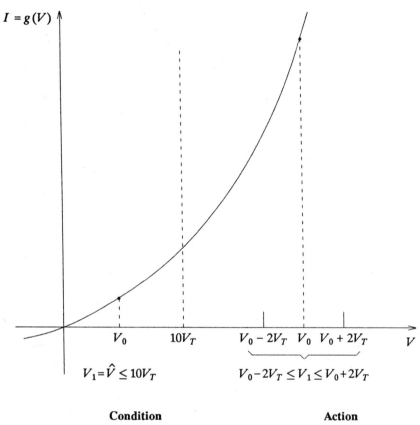

	Condition	Action
1:	$\|\hat{V} - V_0\| \leq 2V_T$	$V_1 = \hat{V}$
2:	$\hat{V} \leq 10V_T$ & $V_0 \leq 10V_T$	$V_1 = \hat{V}$
3:	$\hat{V} < V_0$ & $10V_T < V_0$	$V_1 = V_0 - 2V_T$
4:	$V_0 < \hat{V}$ & $10V_T < \hat{V}$	$V_1 = \max(10V_T, V_0 + 2V_T)$

Figure 4.5 - Two V_T fixed bound junction limiting

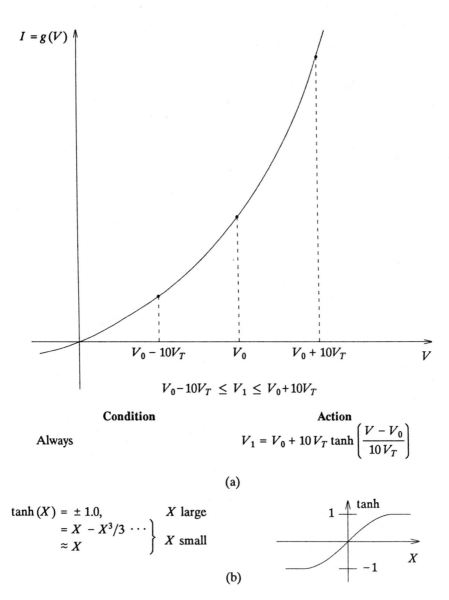

(a)

(b)

Figure 4.6 - (a) Hyperbolic tangent fixed bound junction limiting and (b) hyperbolic tangent function

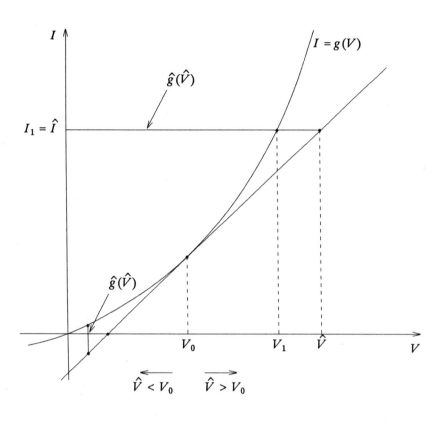

Condition	Action

$\hat{V} \leq 0$ or $\hat{V} \leq V_0$ 　　　　　　　　$V_1 = \hat{V}$

$\hat{V} > 0$ and $\hat{V} > V_0$ 　　　　　　$V_1 = V_0 + V_T \ln\left(\dfrac{\hat{V} - V_0}{V_T} + 1.0 \right)$

Figure 4.7 - Alternating basis auxiliary function junction limiting

$$\hat{g}(V) = \hat{I} \tag{4.23}$$

is used only when $V > V_0$ and $V > 0$. In this case the voltage V_1 at which $g(V)$ and $\hat{g}(V)$ intersect is given by

$$V_1 = V_T \ln\left[\frac{I_1}{I_S} + 1.0\right] \tag{4.24}$$

where

$$I_1 = \hat{g}(\hat{V}) = \hat{I} \tag{4.25}$$

That is, the inverse of the exponential diode equation is used. The actual expression for V_1 given in Figure 4.7 and used in place of (4.24) is easily derived from (4.25) by equating

$$\hat{I} = \frac{I_S}{V_T} e^{\frac{V_0}{V_T}} (\hat{V} - V_0) + I_S (e^{\frac{V_0}{V_T}} - 1.0) \tag{4.26}$$

and

$$I_1 = I_S (e^{\frac{V_1}{V_T}} - 1.0) \tag{4.27}$$

which yields

$$V_1 = V_0 + V_T \ln\left[\frac{\hat{V} - V_0}{V_T} + 1.0\right] \tag{4.28}$$

Examination of (4.28) clearly reveals that voltage limiting is in effect being used; however, a fixed bound is not employed. An additional modification to this method was used in BIAS-3. In the third quadrant, a chord or secant between the point (V_1, I_1) and the origin was used rather than a tangent. This preserved a finite slope and hence a finite conductance rather than a zero conductance thereby assuring that for $V_0 < 0$ and $\hat{V} > 0$, $V_1 > 0$. Model-wise, the chord represented a conductance in parallel with a zero-valued Norton equivalent current source. This latter modification has been found unnecessary in SLIC.

A modified version of the BIAS-3 method as proposed by Colon and implemented by Kao [F11] is summarized in Figure 4.8. As originally proposed, a current iteration is used whenever the diode conductance of the new iterate has a slope larger than a specified value, while a voltage iteration is used otherwise. Since the slope can be uniquely related to a voltage, it is sufficient to define a critical voltage V_{CRIT} rather than compute a slope at each iteration. Nagel [C13] has found that a near optimal choice of V_{CRIT} is the voltage at which the diode equation has a minimum radius of curvature. This voltage is given by

$$V_{CRIT} = V_T \ln(V_T / (\sqrt{2} I_S)) \tag{4.29}$$

Colon's method as modified by Nagel has been used in SPICE-2.

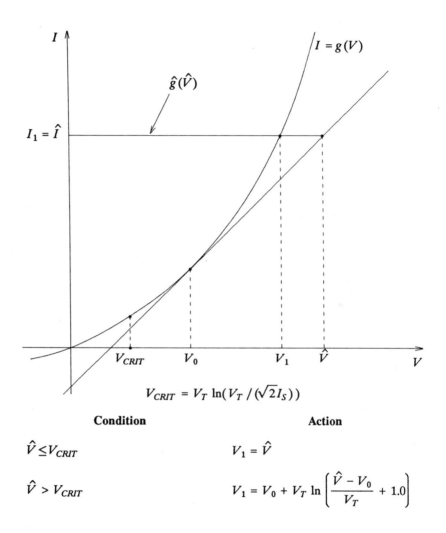

Figure 4.8 - Modified alternating basis method of Colon and Nagel

A method proposed by Cermak [C4] and used in the NICAP program is based on the choice of an auxiliary function given by

$$\hat{g}(V) = -\frac{1}{K}(V - \hat{V}) + \hat{I} \tag{4.30}$$

$$\hat{g}'(V) = -\frac{1}{K} \tag{4.31}$$

as summarized in Figure 4.9. For this method V_1 is defined implicitly by the intersection of $\hat{g}(V)$ and $g(V)$ and hence is the solution of

$$I_S(e^{\frac{V_1}{V_T}} - 1.0) = -\frac{1}{K}(V_1 - \hat{V}) + \hat{I} \tag{4.32}$$

Thus, an iterative solution of (4.32) for V_1 is required. This procedure amounts to Newton-Raphson iteration within Newton-Raphson iteration. The idea is to substitute more inner iterations on scalar equations for fewer outer iterations on matrix equations. Cermak has shown this method to be equivalent to a nonlinear transformation of the original equations into a new space in which exponentials behave linearly.

Two related methods which have been conjectured but apparently not implemented are closely related to Cermak's method and summarized in Figures 4.10 and 4.11. In both cases, $\hat{g}(V)$ is not held constant but rather chosen such that $\hat{g}(V)$ is orthogonal to $g(V)$ at V_1 or V_0 respectively. Again, V_1 is defined implicitly as the solution of

$$I_S(e^{V_1/V_T} - 1) = \frac{-1}{\dfrac{I_S}{V_T}e^{V_1/V_T}}(V_1 - \hat{V}) + \hat{I} \tag{4.33}$$

or

$$I_S(e^{V_1/V_T} - 1) = \frac{-1}{\dfrac{I_S}{V_T}e^{V_0/V_T}}(V_1 - \hat{V}) + \hat{I} \tag{4.34}$$

respectively. Note that the latter approach (4.34) is easier to implement in that $\hat{g}(V)$ is evaluated once at V_0 and does not have to be re-evaluated for each new value of V_1. Finally, note that either of these two variations behaves similarly to the alternating basis techniques for negative or large positive voltages. That is, (4.33) and (4.34) reduce to an iteration on voltage for $\hat{V} < 0$ and to an iteration on current for $\hat{V} \gg 0$.

4.3 Variations of Newton-Raphson Iteration

The Newton-Raphson iteration function in its simplest form was derived in (4.7) for the scalar case. It was pointed out that at a solution $\overline{V}$, $\overline{g(V)} = 0$ and $V^{(n+1)} = V^{(n)}$ as required. This fact does not depend on the exactness with which

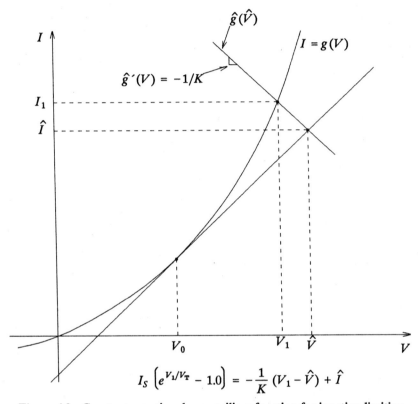

Figure 4.9 - Constant negative slope auxiliary function for junction limiting

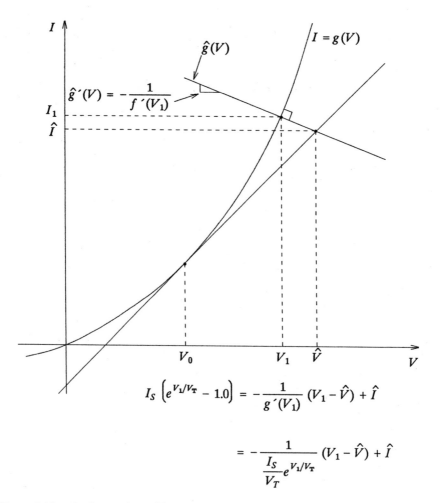

Figure 4.10 - Orthogonal auxiliary function junction limiting where $g'(V) = -1/g'(V_1)$

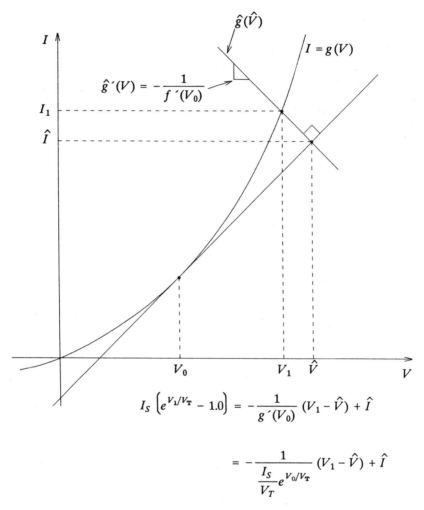

Figure 4.11 - Orthogonal auxiliary function junction limiting where $g'(V) = -1/g'(V_0)$

$\overline{g}'(V^{(n)})$ is computed. This observation suggests that approximations to $\overline{g}'(V^{(n)})$ might be used and, in the extreme, that functions not even resembling a derivative might be used. Several classical variations of Newton-Raphson iteration serve to illustrate the above premise. Before these variations are illustrated, the extension of Newton-Raphson iteration from solution of the classic problem defined by (4.2) to the conventional circuit analysis problem defined by (4.1) should be examined.

The extension of Newton-Raphson iteration from its classical numeric formulation to a conventional circuit formulation is illustrated in Figure 4.12 in three steps. Figure 4.12(a) illustrates the iteration sequence for a classical formulation in seeking the zero of $\overline{g}(V)$ for the simple diode and current source shown. Figure 4.12(b) shows the same iteration sequence for the same circuit where a solution to $g(V) = I$ is sought. Finally, Figure 4.12(c) illustrates the iteration sequence in finding the solution to the conventional "load line" problem where a conductance, G, has been added to the diode-current circuit of Figure 4.12(a) and (b).

This same equivalence can be illustrated in another way as follows: For the circuit of Figure 4.12(c), the classical Newton-Raphson iteration method defined by (4.7) can be applied to find the zeros of

$$\overline{g}(V) = g(V) + G V - I = 0 \tag{4.35}$$

resulting in

$$V^{(n+1)} = V^{(n)} - \overline{g}(V^{(n)})/\overline{g}'(V^{(n)}) \tag{4.36}$$

$$= V^{(n)} - \frac{g(V^{(n)}) + G V^{(n)} - I}{g'(V^{(n)}) + G} \tag{4.37}$$

$$= \frac{I - \left[g(V^{(n)}) - g'(V^{(n)}) V^{(n)}\right]}{g'(V^{(n)}) + G} \tag{4.38}$$

or

$$\left[g'(V^{(n)}) + G\right] V^{(n+1)} = I - \left[g(V^{(n)}) - g'(V^{(n)}) V^{(n)}\right] \tag{4.39}$$

which finally yields

$$(G + G_D{}^{(n)}) V^{(n+1)} = I - I_{DN}^{(n)} \tag{4.40}$$

which is identical to (4.18) based on a more conventional circuit formulation.

For comparison, the Newton-Raphson iteration sequence and three variations are shown in Figure 4.13. The first variation, Figure 4.13(b), is usually referred to as the Chord method where, in essence, a nonlinear function, $g(V)$, is modeled at each iteration by a line of fixed slope (the chord) through the previous solution. The iteration sequence shown is for the same current source-conductance-diode circuit considered previously and is defined by the iteration function

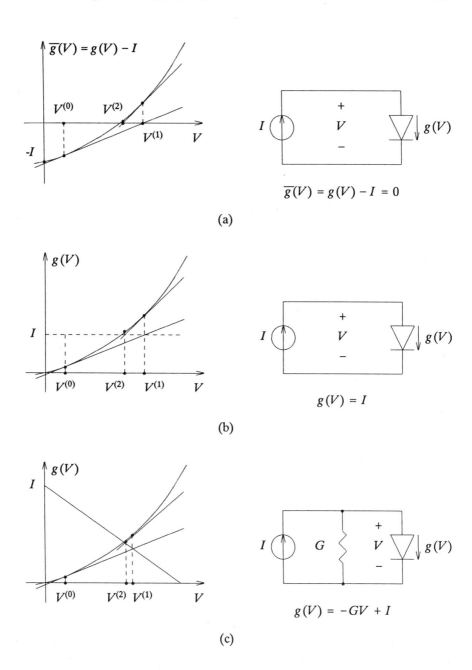

(a)

(b)

(c)

Figure 4.12 - Newton-Raphson iteration for (a) $\overline{g}(V) = 0$, (b) $g(V) = I$, and (c) $g(V) = -GV + I$

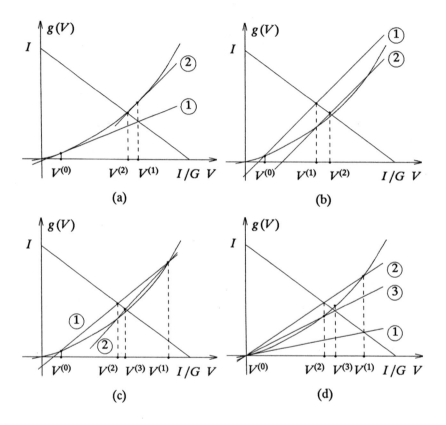

Figure 4.13 - (a) Newton-Raphson iteration, (b) Chord iteration, (c) Secant iteration, and (d) Line-thru-the-origin iteration

$$V^{(n+1)} = V^{(n)} - \overline{g}(V^n)/(M+G)$$ (4.41)

which for $\overline{g}(V^{(n)})$ given by (4.35) can be rewritten in the form

$$(G+M)V^{(n+1)} = I - I_{DN}^{(n)}$$ (4.42)

It should be noted that the slope of the chord in Figure 4.13(b) is given by M. If the iteration sequence of Figure 4.13(b) were redrawn in terms of the derived function $\overline{g}(V)$, the chord would then have to have a slope $M + G$.

In view of the simplicity of the Chord method where no derivative is required, one might question why Newton-Raphson is generally used instead. The answer lies in the rate of convergence. The Chord method can be shown to converge linearly; that is,

$$\left| V^{(n+1)} - V^{(n)} \right| = \varepsilon_C \left| V^{(n)} - V^{(n-1)} \right| \qquad \varepsilon_C < 1$$ (4.43)

while the Newton-Raphson method can be shown to converge quadratically such that

$$\left| v^{(n+1)} - V^{(n)} \right| = \varepsilon_{NR} \left| V^{(n)} - V^{(n-1)} \right|^2$$ (4.44)

In essence, Newton-Raphson can be compared with moving toward a solution in the direction of steepest descent.

The second variation of Newton-Raphson is usually referred to as the Secant method but also known as the method of false position or *Regula Falsi*. It is shown in Figure 4.13(c). In this method, the derivative $\overline{g}'(V)$ of Newton-Raphson is replaced by an approximation

$$\overline{g}'(V^{(n)}) = \frac{\overline{g}(V^{(n)}) - \overline{g}(V^{(n-1)})}{V^{(n)} - V^{(n-1)}}$$ (4.45)

resulting in the iteration function

$$V^{(n+1)} = V^{(n)} - \frac{V^n) - V^{(n-1)}}{\overline{g}(V^{(n)}) - \overline{g}(V^{(n-1)})} \overline{g}(V^{(n)})$$ (4.46)

or, after some rearrangement,

$$V^{(n+1)} = \frac{\overline{g}(V^{(n)}) V^{(n-1)} - \overline{g}(V^{(n-1)}) V^{(n)}}{\overline{g}(V^{(n)}) - \overline{g}(V^{(n-1)})}$$ (4.47)

The Secant method can be shown to be of fractional order, that is, to have a rate of convergence of 1.67. Its potential usefulness is in dealing with nonlinearities in which derivatives are difficult or impossible to evaluate.

The third variation of Newton-Raphson iteration illustrated in Figure 4.13(d) can be viewed as a line-thru-the-origin iteration in which nonlinearities are modeled at each iteration by a simple conductance. This method can be viewed as a Secant method in which one of the two points is always taken as the origin or it can be viewed as a "variable Chord" method in which case M is given by

$$M = \frac{g(V^{(n)})}{V^{(n)}} \tag{4.48}$$

resulting in the iteration function

$$V^{(n+1)} = V^{(n)} - \frac{\overline{g}(V^{(n)})}{G + g(V^{(n)})/V^{(n)}} \tag{4.49}$$

which can be rewritten as

$$(G + g(V^{(n)})/V^{(n)}) V^{(n+1)} = I \tag{4.50}$$

for the same current source-conductance-diode example. Note that $M = g(V^{(n)})/V^{(n)}$ simply represents the diode current divided by the diode voltage at the n th iteration. This procedure is exactly what was used in the third quadrant in the BIAS-3 program as previously pointed out. As a final point, it should be noted that this method is not usually described in numerical analysis textbooks seemingly because its application to the classical problem of solving $\overline{g}(V) = 0$ is not nearly as obvious as its application to solving the conventional circuit problem $g(V) = I$. The latter is particularly true when exponential and square-law nonlinearities typically encountered in electronic circuits problems are considered. It is therefore speculated that this method has a rate of convergence somewhere between that of the Chord and Secant methods, i.e., between 1 and 1.67.

4.4 Internal Device Node Suppression

The final topic to be considered in this chapter on nonlinear equation solution is a technique for suppressing the node equations normally written for internal device nodes. The problem can be illustrated by considering the simple dc models for a junction diode and a bipolar junction transistor with ohmic series resistances and the corresponding indefinite admittance matrices for these elements as shown in Figures 4.14 and 4.15 respectively. It can be seen that for every diode with ohmic losses, at least one additional circuit equation must be written for the internal device node while for every bipolar transistor with ohmic losses, three additional circuit equations must be treated as any other circuit equations--they require storage; they must be processed during Gaussian elimination; they can contribute fill-ins.

The object of the proposed technique is to try to eliminate the explicit formulation of circuit equations for these nodes. The intent is to reach this objective by making use of the fact that built-in device models are characterized by a known topology, advantage of which may be taken. The technique will be illustrated first for the diode and then for the bipolar transistor.

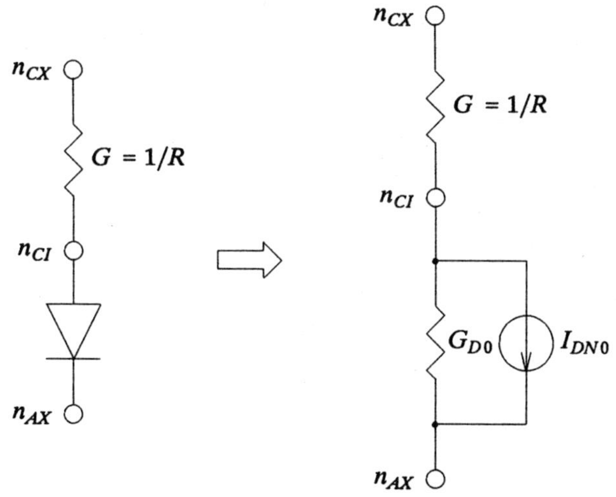

(a)

$$\begin{bmatrix} G & -G & 0 \\ -G & G+G_{D0} & -G_{D0} \\ 0 & -G_{D0} & G_{D0} \end{bmatrix} \begin{bmatrix} V_{CX} \\ V_{CI} \\ V_{AX} \end{bmatrix} = \begin{bmatrix} 0 \\ -I_{DN0} \\ +I_{DN0} \end{bmatrix}$$

(b)

Figure 4.14 - (a) Diode with ohmic resistance and linearized equivalent circuit and (b) indefinite admittance matrix circuit equations

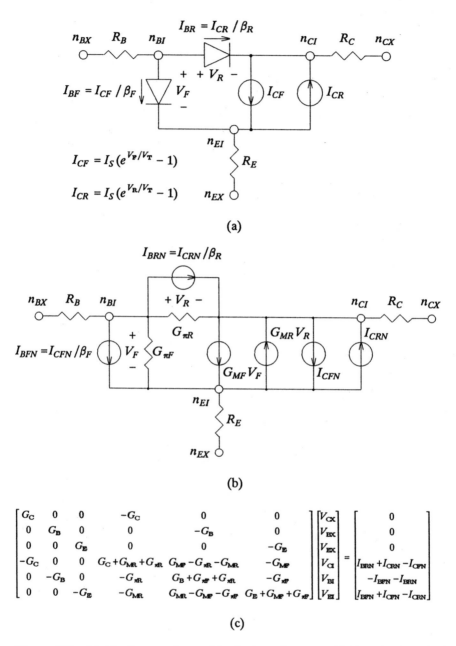

Figure 4.15 - (a) Bipolar transistor with ohmic resistances, (b) linearized equivalent circuit and (c) indefinite nodal equations

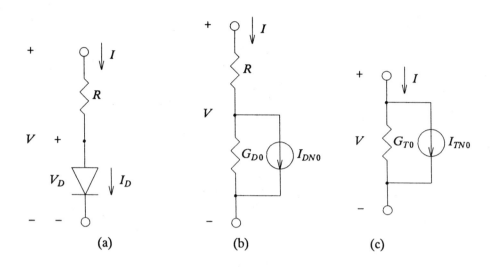

Figure 4.16 - (a) Ideal diode with series resistance; (b) Linearized diode model with series resistance; and (c) Norton equivalent circuit

An ideal diode with series resistance is shown in Figure 4.16(a) and its linearized equivalent circuit is shown in Figure 4.16(b) where as before

$$I_D = I_S(e^{V_D/V_T} - 1) \tag{4.51}$$

$$\approx G_{D0}V_D + (I_{D0} - G_{D0}V_0) \tag{4.52}$$

$$\approx G_{D0}V_D + I_{DN0} \tag{4.53}$$

Viewed as a one-port network, the linearized equivalent circuit of Figure 4.16(b) can itself be modeled as a Norton equivalent circuit as shown in Figure 4.16(c) where

$$G_{T0} = \frac{G_{D0}}{1 + G_{D0}R} \tag{4.54}$$

$$I_{TN0} = \frac{I_{DN0}}{1 + G_{D0}R} \tag{4.55}$$

Here, G_{T0} is the total equivalent conductance seen looking into the one-port with the source I_{DN0} set to zero while I_{TN0} is the total equivalent current which flows into the one-port when terminated by a short circuit. The external current and voltage are thus related by the expression

$$I = G_{T0}V + I_{TN0} \tag{4.56}$$

The conductance G_{T0} and current source I_{TN0} can thus be added to the linearized nodal equations of the complete circuit. Once a new voltage $\hat{V}$ has been established from a solution of the linear equations, a new internal junction voltage, $\hat{V}_D$, can be computed as follows:

$$\hat{I} = G_{T0}\hat{V} + I_{TN0} \tag{4.57}$$

$$\hat{V}_D = \hat{V} - \hat{I}R \tag{4.58}$$

At this point, any of the previous junction limiting algorithms can be used to establish a new trial operating point for the next iteration and the entire process repeated.

In effect, the diode and series resistance have thus been modeled as a single conductance in parallel with a single Norton equivalent current source. Several observations are in order. First, note that for $R = 0$, (4.54) reduces to $G_{T0} = G_{D0}$ as expected. Second, note that in essence, Newton-Raphson has been applied to a modified exponential as indicated in Figure 4.17.

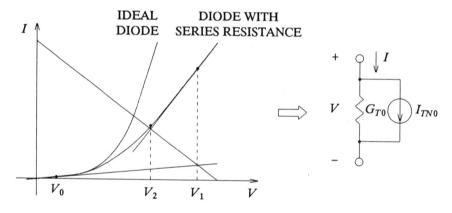

Figure 4.17 - Newton-Raphson with node suppression viewed as iterations on modified exponential

Finally, note that both G_{T0} and I_{TN0} must be saved at each iteration. An attempt at halving this storage price by approximating $\hat{V}_D$ from

$$\hat{V}_D = \hat{V} - I_{D0}R \tag{4.59}$$

where I_{D0} is the saved current from the previous iteration has been found in practice to cause convergence problems.

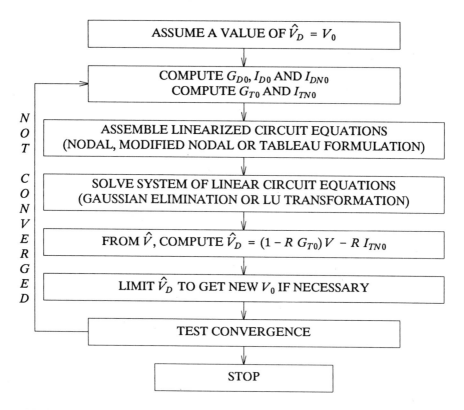

Figure 4.18 - Revised Newton-Raphson algorithm for internal node suppression

The modified Newton-Raphson algorithm corresponding to this approach is given in Figure 4.18.

Because it was a one-port, the diode example considered above lent itself quite easily to localized processing of an internal node. It is apparent from an examination of the non-linear hybrid-pi dc model of a bipolar junction transistor and its linearized equivalent circuit as shown in Figure 4.15(a) and Figure 4.15(b) that because of its multi-port nature, the bipolar transistor will be less easily accommodated than the diode. While several approaches are possible, the one outlined below has thus far proved the most effective. First, the local Norton equivalent current sources are rearranged a shown in Figure 4.19(a). Next, the internal sources and ohmic series resistors are transformed into Thevenin equivalent voltage sources and series resistors as shown in Figure 4.19(b). Here the voltage sources V_{BT}, V_{ET}, and V_{CT} are given by

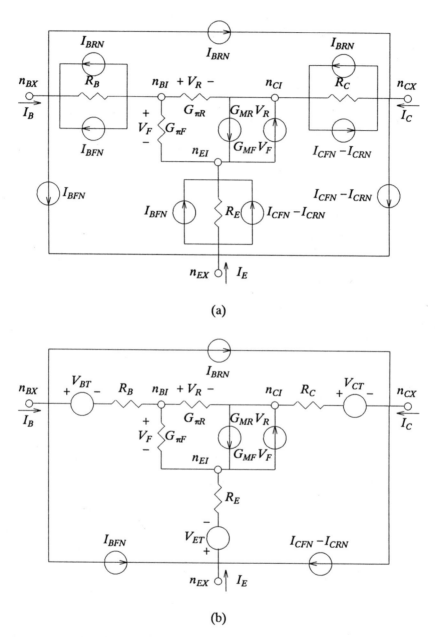

Figure 4.19 - (a) Rearranged Norton equivalent sources and (b) transformation to Thevenin equivalent circuit

$$V_{BT} = (I_{BFN} + I_{BRN}) R_B$$

$$V_{ET} = \left[I_{CRN} - (I_{CFN} + I_{BFN}) \right] R_E \tag{4.60}$$

$$V_{CT} = \left[I_{CFN} - (I_{CRN} + I_{BRN}) \right] R_C$$

The short-circuit y-parameters of the intrinsic hybrid-pi elements with respect to the internal base node can be written by inspection and are given by

$$Y_{EEI} = G_{\pi F} + G_{MF}$$

$$Y_{ECI} = -G_{MR}$$

$$Y_{CEI} = -G_{MF} \tag{4.61}$$

$$Y_{CCI} = G_{\pi R} + G_{MR}$$

The determinant of Y, ΔYI, can thus be evaluated as

$$\Delta Y_I = (G_{\pi F} + G_{MF})(G_{\pi R} + G_{MR}) - G_{MF} G_{MR} \tag{4.62}$$

$$= G_{\pi F} G_{\pi R} + G_{MF} G_{\pi R} + G_{MR} G_{\pi F} \tag{4.63}$$

Given (4.61) and (4.63), the open-circuit z-parameters can be obtained by matrix inversion and are given by

$$Z_{EEI} = (G_{\pi R} + G_{MR})/\Delta Y_I$$

$$Z_{ECI} = G_{MR}/\Delta Y_I$$

$$Z_{CEI} = G_{MF}/\Delta Y_I \tag{4.64}$$

$$Z_{CCI} = (G_{\pi F} + G_{MF})/\Delta Y_I$$

The ohmic series resistances can now be added to the internal z-parameters to obtain external z-parameters given by

$$Z_{EEX} = (G_{\pi R} + G_{MR})/\Delta Y_I + R_B + R_E$$

$$Z_{ECX} = G_{MR}/\Delta Y_I + R_B$$

$$Z_{CEX} = G_{MF}/\Delta Y_I + R_B \tag{4.65}$$

$$Z_{CCX} = (G_{\pi F} + G_{MF})/\Delta Y_I + R_B + R_C$$

and the corresponding determinant, ΔZ_X, evaluated as

$$\Delta Z_X = \left[\frac{G_{\pi F} + G_{MF}}{\Delta Y_I} + R_B + R_C \right] \left[\frac{G_{\pi R} + G_{MR}}{\Delta Y_I} + R_B + R_E \right]$$

$$- \left[\frac{G_{MF}}{\Delta Y_I} + R_B \right] \left[\frac{G_{MR}}{\Delta Y_I} + R_C \right] \tag{4.66}$$

$$= \left[\frac{G_{\pi F}}{\Delta Y_I} + R_C \right] \left[\frac{G_{\pi R}}{\Delta Y_I} + R_E \right] + \left[\frac{G_{MF}}{\Delta Y_I} + R_B \right] \left[\frac{G_{\pi R}}{\Delta Y_I} + R_E \right]$$

$$+ \left[\frac{G_{MR}}{\Delta Y_I} + R_B \right] \left[\frac{G_{\pi F}}{\Delta Y_I} + R_C \right] \tag{4.67}$$

Finally, external y-parameters can be obtained by once again performing a matrix inversion to obtain

$$Y_{EEX} = (\frac{G_{\pi F}}{\Delta Y_I} + G_{MR}/\Delta Y_I + R_B + R_C)/\Delta Z_X$$

$$Y_{ECX} = -(G_{MR}/\Delta Y_I + R_B)/\Delta Z_X$$

$$Y_{CEX} = -(G_{MF}/\Delta Y_I + R_B)/\Delta Z_X \tag{4.68}$$

$$Y_{CCX} = (G_{\pi R}/\Delta Y_I + G_{MR}/\Delta Y_I + R_E)/\Delta Z_X$$

The derived equivalent circuit corresponding to (4.68) is shown in Figure 4.20 where a common base orientation has been used. The currents I_{EI} and I_{CI} are related to the voltages V_{EI} and V_{CI} by

$$I_{EI} = Y_{EEX} V_{EI} + Y_{ECX} V_{CI}$$

$$I_{CI} = Y_{DEX} V_{EI} + Y_{CCS} V_{CI} \tag{4.69}$$

The internal voltages V_{EI} and V_{CI} are related to the externally applied voltages V_{EX} and V_{CI} by the expressions

$$V_{EI} = V_{EX} + V_{BT} - V_{ET}$$

$$V_{CI} = V_{CX} + V_{BT} - V_{CT} \tag{4.70}$$

The equations of (4.69) can now be rewritten in terms of V_{EX} and V_{CS} as

$$I_{EI} = Y_{EEX} V_{EX} + Y_{ECX} V_{CX} + I_{ET}$$

$$I_{CI} = Y_{CEX} V_{EX} + Y_{CCX} V_{CX} + I_{CT} \tag{4.71}$$

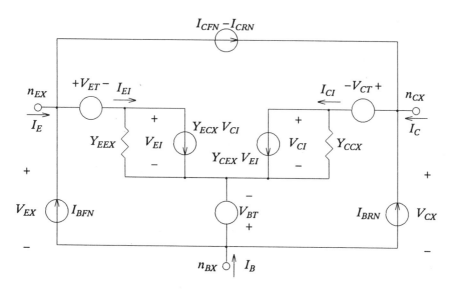

Figure 4.20 - Equivalent circuit of linearized BJT in terms of common-base external y-parameters

where

$$I_{ET} = (Y_{EEX} + Y_{ECX})V_{BT} - Y_{ECX}V_{CT} - Y_{EEX}V_{ET}$$

$$I_{CT} = (Y_{CCX} + Y_{CEX})V_{BT} - Y_{CEX}V_{ET} - Y_{CCX}V_{CT} \tag{4.72}$$

The equivalent circuit of Figure 4.20 can now be simplified to reflect (4.71) and (4.72) as shown in its final form in Figure 4.21.

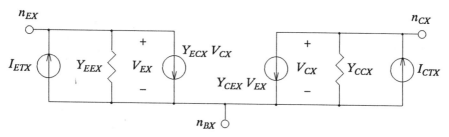

Figure 4.21 - Final Norton equivalent circuit of linearized BJT model in terms of common-base external y-parameters

Here, the source $(I_{CFN} - I_{CRN})$ in Figure 4.20 has been redirected from collector to base and from base to emitter and parallel sources between these nodes have been combined to obtain

$$
\begin{aligned}
I_{ETX} = {} & \left[1 + Y_{ECX}R_B + Y_{EEX}\left(R_B + R_E\right)\right]I_{BFN} \\[2mm]
& + \left[Y_{EEX}R_B + Y_{ECX}\left(R_B + R_C\right)\right]I_{BRN} \\[2mm]
& + \left(1 + Y_{EEX}R_E - Y_{ECX}R_C\right)\left(I_{CFN} - I_{CRN}\right) \\[2mm]
I_{CTX} = {} & \left[1 + Y_{CEX}R_B + Y_{CCX}\left(R_B + R_C\right)\right]I_{BRN} \\[2mm]
& + \left[Y_{CCX}R_B + Y_{CEX}\left(R_B + R_E\right)\right]I_{BFN} \\[2mm]
& + \left(1 + Y_{CCX}R_C - Y_{CEX}R_E\right)\left(I_{CRN} - I_{CFN}\right)
\end{aligned}
\tag{4.73}
$$

where (4.60) has been substituted into (4.72). Thus, the terminal currents can be written as

$$
\begin{aligned}
I_E &= Y_{EEX}V_{EX} + Y_{ECX}V_{CX} + I_{ETX} \\
I_C &= Y_{CEX}V_{EX} + Y_{CCX}V_{CX} + I_{CTX}
\end{aligned}
\tag{4.74}
$$

These results and the equivalent circuit of Figure 4.21 complete the derivation. The essential equations in obtaining the simplified Norton equivalent circuit of Figure 4.21 from the initial Norton equivalent circuit of Figure 4.15(b) are (4.63), (4.67), (4.68) and (4.73). The inverse process of obtaining new internal junction voltage estimates, $\hat{V}_F$ and $\hat{V}_R$, from external voltage estimates $\hat{V}_{EX}$ and $\hat{V}_{CX}$, is based on the evaluation of $\hat{I}_E$ and $\hat{I}_C$ from (4.74). Then

$$
\begin{aligned}
\hat{V}_F &= -\hat{V}_{EI} = -\left(\hat{V}_{EX} - (R_B + R_E)\hat{I}_E - R_B\hat{I}_C\right) \\
\hat{V}_R &= -\hat{V}_{CI} = -\left(\hat{V}_{CX} - (R_B + R_C)\hat{I}_C - R_B\hat{I}_E\right)
\end{aligned}
\tag{4.75}
$$

Again, the linearized bipolar junction transistor with ohmic series resistors has been modeled by a three-terminal (two-port) Norton equivalent circuit. The modified Newton-Raphson algorithm corresponding to this procedure is given in Figure 4.22.

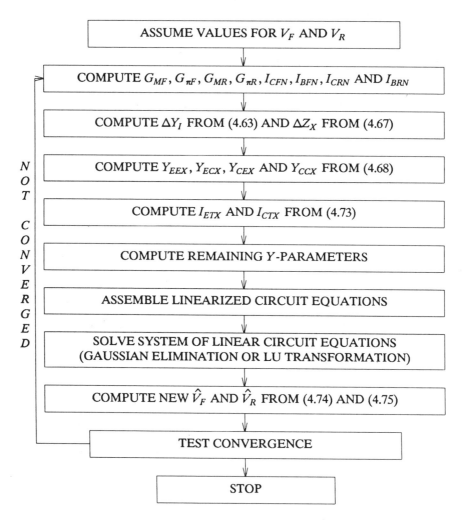

Figure 4.22 - Modified Newton-Raphson algorithm for bipolar transistors

5. NUMERICAL INTEGRATION

5.1 Introduction to Integration Formulas

In general terms, the transient simulation problem can be stated as follows: Given a system of differential-algebraic circuit equations and their solution at times $t_0, \cdots, t_{n-1}, t_n$, it is desired to solve the system of equations for its solution at time t_{n+1}. Assume the system of equations contains capacitors characterized by

$$q = q(V) \tag{5.1}$$

such that

$$I = dq/dt = (dq/dV)(dV/dt) = q'V' \tag{5.2}$$

and/or inductors where

$$\phi = \phi(I) \tag{5.3}$$

such that

$$V = d\phi/dt = (d\phi/dI)(dI/dt) = \phi'I' \tag{5.4}$$

and where the capacitor charges, $q(V)$, and the inductor fluxes, $\phi(I)$ are known functions of voltage and current. Were an algebraic approximation to the derivatives V' and I' available, the system of equations could be converted to a purely algebraic form solvable via Newton-Raphson and/or Gaussian elimination methods.

An example of such an algebraic approximation to a derivative might be

$$X'(t_n) = \frac{X(t_n + h) - X(t_n)}{h} \tag{5.5}$$

or, to simplify notation,

$$X_n' = \frac{X_{n+1} - X_n}{h} \tag{5.6}$$

where $h = (t_{n+1} - t_n)$ will henceforth be referred to as the timestep. Note that (5.6), which will be referred to as a Forward Euler approximation, can be viewed as a difference formula which relates the derivative of X at times t_n (i.e., X_n') to the values of X at times t_{n+1} and t_n (i.e., X_{n+1} and X_n respectively). The specific consequence of approximating derivatives via difference equations is that the true and exact solution of the simulation equations, a continuous function, is at best being approximated only at discrete time points through a process of extrapolation. The practical consequence of such approaches is that different approximations can lead to significantly different results.

By way of example, consider the effect of a constant voltage source E applied to a series RC circuit as shown in Figure 5.1(a). In terms of the capacitor voltage, the circuit equation may be written in the form

$$E = \tau V' + V \tag{5.7}$$

where $\tau = RC$. A slight rearrangement leads to

$$V' = (E - V)/\tau \tag{5.8}$$

$$= f(V) \tag{5.9}$$

A rearrangement of the Forward Euler approximation, (5.6), results in

$$V_{n+1} = h V_n' + V_n \tag{5.10}$$

or, with the substitution of (5.9), in

$$V_{n+1} = h f(V_n) + V_n \tag{5.11}$$

A physical interpretation is illustrated in Figure 5.1(b). It can easily be seen that the selection of an arbitrarily large timestep can lead to a significant error.

Suppose that a different approximation to a derivative such as

$$X_{n+1}' = \frac{X_{n+1} - X_n}{h} \tag{5.12}$$

is employed. Known as the Backward Euler approximation, (5.12) relates the derivative at time t_{n+1} to the values of X at times t_{n+1} and t_n. Its application to the example just considered leads to

$$V_{n+1} = h V_{n+1}' + V_n \tag{5.13}$$

$$= h f(V_{n+1}) + V_n \tag{5.14}$$

and is illustrated in Figure 5.1(c). It can be seen that for arbitrarily large choices of step-size, the latter approximation will result in a computed solution which more nearly approximates the exact solution for this problem than does the Forward Euler method. This difference between the Forward and Backward Euler methods is a manifestation of the stability properties of the two methods. Stability becomes of significant importance when large timesteps are considered.

The above example leads to several other observations. First, it can be pointed out that as the timestep h is diminished, (5.12) and (5.6) converge. Further the similarity of (5.10) to a truncated Taylor series expansion about t_n suggests that the error in evaluating V_{n+1} by either method will be proportional to h^2 multiplied by the second derivative of $V(t)$ evaluated someplace between t_n and $t_n + h$. This error is referred to as truncation error and represents a localized property of a method.

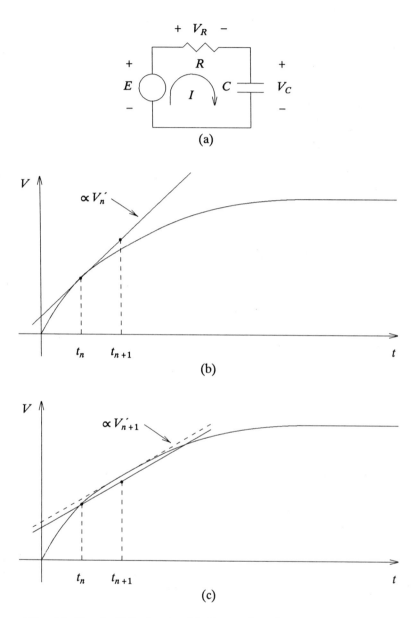

Figure 5.1 - (a) Simple RC circuit, (b) Forward Euler and (c) Backward Euler
approximations applied to RC step response

Next, the analogy between the Forward Euler approximation as written in (5.10) and a Taylor series leads to speculation that more accurate approximations (i.e., smaller truncation error) than (5.6) and (5.12) exist. This is in fact the case and leads to a consideration of the order of a method or approximation as an indicator of truncation error. Higher order methods possess error terms proportional to higher-order derivatives which like high-order terms in a Taylor series tend to vanish. Both the Forward and Backward Euler methods are examples of first-order methods. An example of a second-order method is the well known Trapezoidal formula

$$X_{n+1} = X_n + \frac{h}{2} (X_{n+1}' + X_n')$$
(5.15)

which implies that one extrapolate from the present value, X_n, to the new value, X_{n+1}, along a path whose slope is given by the average of the derivatives at $t_n + h$ and t_n. For the example above, the Trapezoidal formula is illustrated in Figure 5.2.

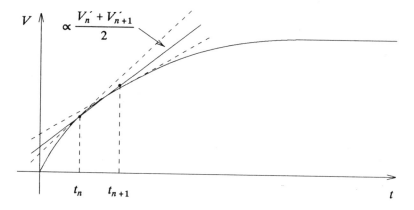

Figure 5.2 - The second-order Trapezoidal approximation applied to an RC step response

Again, comparing the Forward and Backward Euler methods as written in (5.11) and (5.14) respectively, it can be seen that V_{n+1} in (5.11) can be evaluated explicitly in terms of the known value V_n at time t_n and the known function $f(V)$. In contrast, (5.14) represents an implicit equation in the unknown V_{n+1}. The Trapezoidal formula (5.15) also represents an implicit method, which, in general, is characterized by the appearance of a term involving the derivative at the new time point $t_n + h$ (i.e., x_{n+1}'). The significance of implicit methods is that they usually possess superior stability properties over explicit methods.

For the example above of a linear circuit, an implicit method is no more difficult to apply than an explicit method. However, if $f(V)$ were a nonlinear function, the use of an implicit method would necessitate finding the solution of a nonlinear algebraic equation by numerical methods, such as Newton-Raphson iteration. In an attempt to avoid such computation, an approach sometimes employed has been to use an explicit method to predict a first guess at the new value and use that prediction in an implicit method to obtain a refinement or correction. This predictor-corrector technique does not derive the full benefit of the stability properties of implicit methods and consequently has not proved as useful in circuit simulation programs. The reason is that less computer time is required to solve circuit equations at a few widely spaced time points even using Newton-Raphson iteration than is required to solve circuit equations explicitly at the many time points required by explicit or predictor-corrector methods.

Before concluding this section, consider another second-order formula,

$$X_{n+1} = \frac{4}{3} X_n - \frac{1}{3} X_{n-1} + \frac{2}{3} h X'_{n+1} \tag{5.16}$$

which will be referred to as the Gear-Shichman formula. Previously considered formulas included values and derivatives at t_n and hence were single-step formulas, (5.16) also includes a value at $t_n - h$ and hence is a two-step formula. In general, multistep formulas do exist, can be implicit or explicit, and can be of different orders.

As they will prove useful in presenting subsequent topics, the four integration formulas introduced thus far, Forward Euler in (5.6), Backward Euler in (5.12), Trapezoidal in (5.15), and Gear-Shichman in (5.16), are summarized below in Table 5.1. Note that the formulas are expressed both in terms of the value and in terms of the derivative at time $t_n + h$ (i.e., X_{n+1} and X'_{n+1} respectively).

FE	$X_{n+1} = X_n + h X'_n$	$X'_n = \dfrac{X_{n+1} - X_n}{h}$
BE	$X_{n+1} = X_n + h X'_{n+1}$	$X'_{n+1} = \dfrac{X_{n+1} - X_n}{h}$
TR	$X_{n+1} = X_n + \dfrac{h}{2} (X'_{n+1} + X'_n)$	$X'_{n+1} = 2 \left(\dfrac{X_{n+1} - X_n}{h} \right) - X'_n$
GS	$X_{n+1} = \dfrac{4}{3} X_n - \dfrac{1}{3} X_{n-1} + \dfrac{2h}{3} X'_{n+1}$	$X'_{n+1} = \dfrac{3}{2h} X_{n+1} - \dfrac{2}{h} X_n + \dfrac{1}{2h} X_{n-1}$

TABLE 5.1 - Summary of Forward Euler (FE), Backward Euler (BE), Trapezoidal (TR), and Gear-Shichman (GS) Integration Formulas

5.2 Application of Integration Formulas

If the branch equations characterizing capacitors and inductors, (5.2) and (5.4) respectively, are rewritten in the same notation as (5.6) at time $t_n + h$, the result is

$$I_{n+1} = q'_{n+1}V'_{n+1} \qquad (5.17)$$

and

$$V_{n+1} = \phi'_{n+1}I'_{n+1} \qquad (5.18)$$

respectively. It can be seen that what is required is not the derivatives at time t_n but rather the derivatives at time t_{n+1}. An implicit method is called for. This fact leads to the consideration of the class of difference formulas known as backward difference formulas (BDF's) of the form

$$X'_{n+1} = \frac{1}{h}\sum_{i=0}^{k}\alpha_i X_{n+1\text{-}i} \qquad (5.19)$$

which expresses the derivative at $t_n + h$ in terms of the function value at $t_n + h$ as well as previous function values. Note that the Backward Euler and Gear-Shichman formulas, (5.12) and (5.16) respectively, as can be seen in Table 5.1, can be put into the form of (5.19) and are thus seen to be BDF's. The Trapezoidal formula (5.15) which contains X'_n as well as X'_{n+1} is not, strictly speaking, a BDF but can be used in the same way as will now be brought out with several examples.

Though the methods were presented differently, the program ECAP [C2] employed Backward Euler integration for capacitors and Trapezoidal integration for inductors and, as such, was probably the first circuit simulation program to make use of implicit integration methods. Consider the case of a linear capacitor where (5.17) becomes

$$I_{n+1} = C V'_{n+1} \qquad (5.20)$$

such that from (5.12)

$$I_{n+1} = \frac{C}{h}(V_{n+1} - V_n) \qquad (5.21)$$

$$= \frac{C}{h}V_{n+1} - \frac{C}{h}V_n \qquad (5.22)$$

$$= G_{Ceq}V_{n+1} + I_{Ceq} \qquad (5.23)$$

where $G_{Ceq} = C/h$ and $I_{Ceq} = -(C/h)V_n$. Next, consider a linear inductor where (5.18) becomes

$$V_{n+1} = L I'_{n+1} \qquad (5.24)$$

From (5.15) and (5.24)

$$I_{n+1} = \frac{h}{2}(I'_{n+1} + I'_n) + I_n \tag{5.25}$$

$$= \frac{h}{2}(\frac{V_{n+1}}{L} + \frac{V_n}{L}) + I_n \tag{5.26}$$

$$= \frac{h}{2L}V_{n+1} + \frac{h}{2L}V_n + I_n \tag{5.27}$$

$$= G_{Leq}V_{n+1} + I_{Leq} \tag{5.28}$$

where $G_{Leq} = h/2L$ and $I_{Leq} = (h/2L)V_n + I_n$. Thus from (5.23) and (5.28), it can be seen that either element can easily be included in a nodal analysis formulation of circuit equations.

Similarly, the Gear-Shichman formula, (5.16), can also be used and gives

$$I_{n+1} = \frac{3}{2}\frac{C}{h}V_{n+1} - \frac{2C}{h}V_n + \frac{C}{2h}V_{n-1} \tag{5.29}$$

and

$$I_{n+1} = \frac{2}{3}\frac{h}{L}V_{n+1} + \frac{4}{3}I_n - \frac{1}{3}I_{n-1} \tag{5.30}$$

for the capacitor and inductor respectively. These results are summarized in Table 5.2. Note that for this type of treatment the Forward Euler formula is difficult to apply in that it does not involve a derivative term at time $t_n + h$.

The same approach can be extended to nonlinear elements. For example, consider a semiconductor diode, where ohmic losses are ignored for convenience. The total current through the diode is given by

$$I_T = I_V + I_\tau + I_q \tag{5.31}$$

$$= I_V + \tau\frac{dI_V}{dt} + \frac{dq(V)}{dt} \tag{5.32}$$

where

$$I_V = I_S(e^{V/V_T} - 1) \tag{5.33}$$

and τ is the diode time constant while $q(V)$ is the charge associated with the depletion region. From (5.33)

$$I_{V_{n+1}} = I_S(e^{V_{n+1}/V_T} - 1) \tag{5.34}$$

For the term I_τ the use of the Gear-Shichman formula results in

$$I_{\tau_{n+1}} = \tau\frac{dI_{V_{n+1}}}{dt} = \frac{3\tau}{2h}I_{V_{n+1}} - \frac{2\tau}{h}I_{V_n} + \frac{\tau}{2h}I_{V_{n-1}} \tag{5.35}$$

Similarly, if

Linear Capacitor $(I = C V')$			
Method	Formula	G_{eq}	I_{eq}
FE	$I_n = C V_n'$	(see text)	(see text)
BE	$I_{n+1} = \dfrac{C}{h} V_{n+1} - \dfrac{C}{h} V_n$	$\dfrac{C}{h}$	$\dfrac{C}{h} V_n$
TR	$I_{n+1} = \dfrac{2C}{h} V_{n+1} - \dfrac{2C}{h} V_n - V_n'$	$\dfrac{2C}{h}$	$-\dfrac{2C}{h} V_n - V_n'$
GS	$I_{n+1} = \dfrac{3C}{2h} V_{n+1} - \dfrac{2C}{h} V_n + \dfrac{C}{2h} V_{n-1}$	$\dfrac{3C}{2h}$	$-\dfrac{2C}{h} V_n + \dfrac{C}{2h} V_{n-1}$

Linear Inductor $(V = L I')$			
Method	Formula	G_{eq}	I_{eq}
FE	$I_{n+1} = \dfrac{h}{L} V_n + I_n$		$\dfrac{h}{L} V_n + I_n$
BE	$I_{n+1} = \dfrac{h}{L} V_{n+1} + I_n$	$\dfrac{h}{L}$	I_n
TR	$I_{n+1} = \dfrac{h}{2L} V_{n+1} + \dfrac{h}{2L} V_n + I_n$	$\dfrac{h}{2L}$	$\dfrac{h}{2L} V_n + I_n$
GS	$I_{n+1} = \dfrac{2h}{3L} V_{n+1} + \dfrac{4}{3} I_n - \dfrac{1}{3} I_{n-1}$	$\dfrac{2h}{3L}$	$\dfrac{4}{3} I_n - \dfrac{1}{3} I_{n-1}$

TABLE 5.2 - Application of Forward Euler (FE), Backward Euler (BE), Trapezoidal (TR), and Gear-Shichman (GS) Integration Formulas to Linear Capacitor and Linear Inductor

$$q = q(V) = \frac{C_0 \phi}{m-1} \left(1 - \frac{V}{\phi}\right)^{1-m} \tag{5.36}$$

such that

$$C = q'(V) = \frac{C_0 \phi}{(m-1)} (1-m)\left(1 - \frac{V}{\phi}\right)^{-m} \left(-\frac{1}{\phi}\right) = \frac{C_0}{\left(1 - \frac{V}{\phi}\right)^m} \tag{5.37}$$

then the use of the Gear-Shichman formula leads to

$$q_{n+1} = \frac{2}{3} h \, q'_{n+1} + \frac{4}{3} q_n - \frac{1}{3} q_{n-1} \tag{5.38}$$

$$= \frac{2}{3} h \, I_{q_{n+1}} + \frac{4}{3} q_n - \frac{1}{3} q_{n-1} \tag{5.39}$$

such that

$$I_{q_{n+1}} = \frac{dq(V_{n+1})}{dt} = \frac{3}{2h} q_{n+1} - \frac{2}{h} q_n + \frac{1}{2h} q_{n-1} \tag{5.40}$$

Because an implicit integration formula was used, the current components $I_{V_{n+1}}$, $I_{\tau_{n+1}}$ and $I_{q_{n+1}}$ are all implicit functions of the voltage V_{n+1}. Consequently, Newton-Raphson iteration, or its equivalent must be used to solve for V_{n+1}. If the iterate values are denoted as $V_{n+1}^{(0)}, \dots V_{n+1}^{(\mu-1)}, V_{n+1}^{(\mu)}, \dots$, the terms involving V_{n+1} can be linearized via a Taylor series expansion which results in the following expressions:

$$I_{V_{n+1}}^{(\nu)} = G_{n+1}^{(\mu-1)} V_{n+1}^{(\nu)} + (I_{V_{n+1}}^{(\mu-1)} - G_{n+1}^{(\mu-1)} V_{n+1}^{(\mu-1)})$$

$$I_{\tau_{n+1}}^{(\nu)} = \frac{3\tau}{2h} G_{n+1}^{(\mu-1)} V_{n+1}^{(\nu)} + \frac{3\tau}{2h} (I_{V_{n+1}}^{(\mu-1)} - G_{n+1}^{(\mu-1)} V_{n+1}^{(\mu-1)}) - \frac{2\tau}{h} I_{V_n} + \frac{\tau}{2h} I_{V_{n-1}} \tag{5.41}$$

$$I_{q_{n+1}}^{(\nu)} = \frac{3 C_{n+1}^{(\mu-1)}}{2h} V_{n+1}^{(\nu)} + \frac{3}{2h} (q_{n+1}^{(\mu-1)} - C_{n+1}^{(\mu-1)} V_{n+1}^{(\mu-1)}) - \frac{2}{h} q_n + \frac{1}{2h} q_{n-1}$$

where

$$G_{n+1}^{(\mu-1)} = \frac{I_S}{V_T} e^{V_{n+1}^{(\mu-1)}/V_T}$$

$$I_{V_{n+1}}^{(\mu-1)} = I_S \left(e^{V_{n+1}^{(\mu-1)}/V_T} - 1 \right)$$

$$\tag{5.42}$$

$$C_{n+1}^{(\mu-1)} = C_0 \left(1 - \frac{V_{n+1}^{(\mu-1)}}{\phi}\right)^{-m}$$

$$q_{n+1}^{(\mu-1)} = \frac{C_0 \phi}{m-1} \left(1 - \frac{V_{n+1}^{(\mu-1)}}{\phi}\right)^{1-m}$$

while

$$I_{V_n} = I_S \, (e^{V_n/V_T} - 1)$$

$$I_{V_{n-1}} = I_S \, (e^{V_{n-1}/V_T} - 1)$$

$$q_n = \frac{C_0 \phi}{m-1} \, (1 - \frac{V_n}{\phi})^{1-m}$$

(5.43)

$$q_{n-1} = \frac{C_0 \phi}{m-1} \, (1 - \frac{V_{n-1}}{\phi})^{1-m}$$

Thus from (5.41), I_T can be written

$$I_{T_{n+1}}^{(v)} = (G_{n+1}^{(\mu-1)} + \frac{3\tau}{2h} G_{n+1}^{(\mu-1)} + \frac{3 C_{n+1}^{(\mu-1)}}{2h}) V_{n+1}^{(v)}$$

$$+ (I_{V_{n+1}}^{(\mu-1)} - G_{n+1}^{(\mu-1)} V_{n+1}^{(\mu-1)})$$

$$+ \frac{3\tau}{2h} (I_{V_{n+1}}^{(\mu-1)} - G_{n+1}^{(\mu-1)} V_{n+1}^{(\mu-1)}) - \frac{2\tau}{h} I_{V_n} + \frac{\tau}{2h} I_{V_{n-1}}$$

$$+ \frac{3}{2h} (q_{n+1}^{(\mu-1)} - C_{n+1}^{(\mu-1)} V_{n+1}^{(\mu-1)}) - \frac{2}{h} q_n + \frac{1}{2h} q_{n-1}$$

(5.44)

$$= G_{T_{eq}}^{(\mu-1)} V_{n+1}^{(v)} + I_{T_{eq}}^{(\mu-1)}$$

(5.45)

Note that in the treatment above, charge has been considered as the fundamental variable associated with the depletion region rather than capacitance. This choice of variables is predicated on the argument developed and presented by Calahan [H12] to the effect that charge leads to a more stable integration.

5.3 Construction of Integration Formulas

To this point, no indication has been given as to how numerical integration formulas are derived or, more properly, what constraints, if any, they must satisfy. The general form for a numerical integration formula is given by

$$X_{n+1} = \sum_{i=0}^{k} a_i X_{n-i} + h \sum_{j=-1}^{l} b_j X_{n-j}'$$

(5.46)

It can be seen that each of the formulas (5.6), (5.12), (5.15) and (5.16) represent specific cases of (5.46) for particular choices of the a_i's and b_j's. In general, formulas for which $b_{-1} = 0$ such as (5.6) represent explicit integration methods. Formulas for which $b_{-1} \neq 0$ represent implicit formulas while formulas for which only $b_{-1} \neq 0$ represent the special case of implicit integration formulas referred to as backward difference formulas (BDF's).

Based on the conjecture that analysis is usually easier than synthesis, it should be easier to demonstrate that the Gear-Shichman formula is second-order and then to show how the coefficients were derived.

The order of an integration formula is considered to be the order of the polynomial (arbitrary) of largest degree for which the formula yields an exact solution. Thus for a polynomial

$$p(t) = \sum_{i=0}^{k} C_i t^i \tag{5.47}$$

the Gear-Shichman formula, (5.16), leads to

$$p(t_{n+1}) = \frac{2}{3} h \, p'(t_{n+1}) + \frac{4}{3} p(t_n) - \frac{1}{3} p(t_{n-1}) \tag{5.48}$$

or

$$\sum_{i=0}^{k} C_i t_{n+1}^i = \frac{2}{3} h \sum_{i=1}^{k} i \, C_i t_n^{i-1} + \frac{4}{3} \sum_{i=0}^{k} C_i t_{n-1}^i - \frac{1}{3} \sum_{i=0}^{k} C_i t_{n-2}^i \tag{5.49}$$

Since $t_{n+1} = t_n + h$ and $t_{n-1} = t_n - h$, (5.49) can be written

$$\sum_{i=0}^{k} C_i (t_n + h)^i = \frac{2}{3} h \sum_{i=1}^{k} i \, C_i (t_n + h)^{i-1} + \frac{4}{3} \sum_{i=0}^{k} C_i t_n^i - \frac{1}{3} \sum_{i=0}^{k} C_i (t_n - h)^i \tag{5.50}$$

In order for the Gear-Shichman formula to result in an exact solution for (5.47) up to order k, the collection of terms for each power of t_n up to and including k must identically vanish. Thus for $k = 0$, (5.50) reduces to

$$C_0 = \frac{4}{3} C_0 - \frac{1}{3} C_0 = C_0 \tag{5.51}$$

such that the formula is at least of order zero. If $k = 1$ is skipped and $k = 2$ is considered, the collection of terms in t^0 (i.e., constant terms) results in

$$C_2 h^2 + C_1 h + C_0 = \frac{2}{3} h \, (2C_2 h + C_1) + \frac{4}{3} C_0 - \frac{1}{3} C_2 h^2 + \frac{1}{3} C_1 h - \frac{1}{3} C_0$$

$$= C_2 h^2 + C_1 h + C_0 \tag{5.52}$$

If terms in t^1 are collected next, the result is

$$2C_2h + C_1 = \frac{2}{3}h\,(2C_2) + \frac{4}{3}C_1 - \frac{1}{3}(-2h\,C_2 + C_1)$$

$$= 2C_2h + C_1 \tag{5.53}$$

Finally, collecting terms in t^2 results in

$$C_2 = \frac{2}{3}h\,(0) + \frac{4}{3}C_2 - \frac{1}{3}C_2 = C_2 \tag{5.54}$$

Thus it can be concluded that the formula is of order two. To see that it is not of order three, one need only repeat the process followed above for $k = 3$ and thereby discover that an exact solution is unattainable.

The process of synthesis or the determination of coefficients of (5.46) for a numerical integration formula of order k can be closely related to the analysis process followed above. An arbitrary polynomial of degree k such as (5.47) can be substituted into (5.46). Terms of like degree in t can be equated. The resulting set of simultaneous equations in terms of a_i's and b_j's can then be solved for the desired coefficients. In effect an exact solution for a single arbitrary polynomial of order k has been matched at k time points.

An alternate synthesis procedure of simpler implementation results by insisting that the integration formula be exact for k special polynomials of the form

$$p(t) = t^m \qquad m = 0, 1, 2 \ldots, k \tag{5.55}$$

at a single arbitrary time point such as $t = 0$. Consider a formula of the Gear-Shichman form

$$X_{n+1} = h\,b_{-1}X_{n+1}' + a_0 X_n + a_1 X_{n-1} \tag{5.56}$$

which, when applied to (5.55), results in

$$t_{n+1}^m = m\,h\,b_{-1}t_{n+1}^{m-1} + a_0 t_n^m + a_1 t_{n-1}^m \tag{5.57}$$

$$= m\,h\,b_{-1}t_{n+1}^{m-1} + a_0(t_{n+1}-h)^m + a_1(t_{n+1}-2h)^m \tag{5.58}$$

To find the coefficients for a second-order formula $(k = 2)$, let $m = 0,1,2$ respectively at $t_{n+1} = 0$ which yields

$$1 = a_0 + a_1 \qquad\qquad (m = 0)$$

$$0 = h\,b_{-1} + a_0(-h) + a_1(-2h) \qquad (m = 1) \tag{5.59}$$

$$0 = a_0(-h)^2 + a_1(-2h)^2 \qquad (m = 2)$$

The solution of (5.59) is easily found to be $b_{-1} = 2/3$, $a_0 = 4/3$ and $a_1 = -1/3$ which are the coefficients of the Gear-Shichman formula.

The synthesis process can be summarized as follows: A difference equation for an arbitrary polynomial of degree k is constructed and forced to be exact at k time points or a series of difference equations are constructed for special polynomials of degree $0,1,2,...,k$ of the form t^m and are forced to be exact at time $t = 0$. Note that the latter approach derives its validity from the fact that any arbitrary polynomial of degree k can be derived as a linear combination of the special polynomials.

Methods which are exact for linear polynomials, that is for $m = 0$ and $m = 1$ are said to be *consistent*. Consistency is a desirable property and is important in formally relating the additional properties of convergence and stability to each other. Loosely speaking, convergence implies that, a sufficiently small timestep h can be chosen to produce an arbitrary degree of accuracy in the computed solution. Stability will be treated in a subsequent section where the consistency requirement will be used to determine which of several possible roots of a characteristic difference equation is in fact the principle root.

The only strict requirement in the above process is that the number of a_i's and b_j's be greater than the desired order. If there are one or more extra coefficients, they may be used as free parameters and selected to improve error control and/or stability. For example, Liniger and Willoughby[H10] previously proposed use of the formula

$$X_{n+1} = h [(1-\mu)X_{n+1}' + \mu X_n'] + x_n \qquad 0 \leq \mu \leq \frac{1}{2} \qquad (5.60)$$

it can be seen that for $\mu=0$, the Backward Euler formula results while for $\mu=1/2$ the Trapezoidal formula is obtained. As will be brought out later, the former formula has a larger region of stability while the latter formula has a smaller truncation error. The free parameter is adjusted to result in the best compromise within the context of the particular problem being solved. Finally, the choice of which a_i's and b_j's to make nonzero has traditionally been the center of debate on error control and stability.

To this point, it has been assumed that a constant timestep h is used. In many cases where it is recognized that voltages and currents are not changing rapidly with time, it may be desirable to increase the timestep h and thereby decrease the total required computer time. As the Backward Euler and Trapezoidal formulas require only a single time interval (i.e., $h = t_{n+1}-t_n$), there is no problem changing timesteps. The Gear-Shichman formula, however, makes use of two time intervals (i.e., $h_1 = t_{n+1}-t_n$ and $h_2 = t_n-t_{n-1}$). Consequently, the formula given by (5.16) is no longer correct when the two timesteps are unequal. A more general formula can easily be obtained from (5.57) by recognizing that $t_{n-1} = t_{n+1}-h_1-h_2$ such that (5.59) can be rewritten as

$$1 = a_0 + a_1 \qquad\qquad (m = 0)$$

$$0 = h\, b_{-1} + a_0(-h_1) + a_1(-h_1-h_2) \quad (m = 1) \tag{5.61}$$

$$0 = a_0(-h_1)^2 + a_1(-h_1-h_2)^2 \qquad (m = 2)$$

which results in the following formula used by Shichman [H11]

$$X_{n+1}' = \frac{h_1(h_1+h_2)}{2h_1+h_2}X_{n+1}' + \frac{(h_1+h_2)^2}{h_2(2h_1+h_2)}X_n - \frac{h_1^2}{h_2(2h_1+h_2)}X_{n-1} \tag{5.62}$$

Note that (5.62) reduces to (5.16) for $h = h_1 = h_2$.

While truncation error and stability are considered in more detail in later sections, it is convenient to illustrate the development of variable-order, variable timestep formulas at this point. Such formulas have been recommended for use in maximizing timesteps concomitant with maintaining stability and a tolerable degree of truncation error. Gear [H5], Brayton, Hachtel and Gustavson [H15], and more recently von Bokhoven [H19] have all considered the special case of backward difference formulas (BDF's) which can be expressed in the general form

$$X_{n+1}' = \sum_{i=0}^{k} \alpha_i X_{n+1-i} \tag{5.63}$$

Again, consideration of the special polynomials, (5.55), at $t = t_{n+1} = 0$ leads to a system of equations of the form

$$
\begin{bmatrix} 0 \\ 1 \\ 0 \\ \cdot \\ 0 \end{bmatrix}
=
\begin{bmatrix}
1 & 1 & 1 & \cdots & 1 \\
(-0) & (-h_1) & (-h_1-h_2) & & (-h_1-h_2...-h_k) \\
(-0)^2 & (-h_1)^2 & (-h_1-h_2)^2 & & (-h_1-h_2...-h_k)^2 \\
\cdot & \cdot & \cdot & & \cdot \\
(-0)^k & (-h_1)^k & (-h_1-h_2)^k & & (-h_1-h_2...-h_k)^k
\end{bmatrix}
\begin{bmatrix} \alpha_0 \\ \alpha_1 \\ \alpha_2 \\ \cdot \\ \alpha_k \end{bmatrix}
\tag{5.64}
$$

The square matrix in (5.64) is of a special type known as a van der Monde matrix. Among its properties is the fact that its determinant can be written explicitly as

$$0 = (-h_1)(-h_2-h_2)(-h_2)(-h_1-h_2-h_3)(-h_2-h_3)(-h_3)...(-h_k) \tag{5.65}$$

This and related properties can be used to derive closed form solutions for $\alpha_0, \alpha_1, ..., \alpha_k$ and to derive new values from the old values at each new time point. For the special case of $h = h_1 = \cdots = h_k$, the resulting coefficients are those derived by Gear [H5]. Brayton, et. al. [H15] considered (in slightly modified form) the general case and demonstrated an equivalence to solving for Lagrange interpolating polynomials, a process which also makes use of the inverse van der Monde matrix. Finally, von Bokhoven [H19] has modified the latter approach by applying Newton's divided difference interpolation polynomials rather than Lagrange polynomials. This modification results in an equivalent closed form solution although intermediate steps appear quite different. Instead of using the previously calculated function values themselves, predictions extrapolated from these values are derived recursively and used to establish and evaluate the BDF. Greater overall efficiency is claimed as a result.

5.4 Truncation Error of Integration Formulas

Truncation error and stability concepts can be illustrated effectively by means of elementary circuit example. Again, consider a series RC network excited at time $t = 0$ by a voltage source E. In this case, assume the voltage across the resistor, $V_R(t)$, is of interest. At time t_n the circuit equation can be written as

$$V_R(t_n) = R\,I(t_n) = R\,C\,\frac{d}{dt}V_C(t_n) \tag{5.66}$$

where $I(t_n)$ is the loop current and $V_C(t_n)$ is the voltage across the capacitor.

If the voltage source is assumed to be a step function of strength E_0 applied at time $t = 0$ and therefore constant for time $t_n > 0$, then

$$V_R(t_n) = R\,C\,\frac{d}{dt}(E_0 - V_R(t_n)) = -R\,C\,\frac{d}{dt}V_R(t_n) \tag{5.67}$$

As before, let $\tau = R\,C$ such that

$$V_R(t_n) = -\tau V_R'(t_n) \tag{5.68}$$

or

$$V_R'(t_n) = \lambda V_R(t_n) \tag{5.69}$$

where $\lambda = -1/\tau$.

If the derivative forms of Forward Euler (FE), Backward Euler (BE), Trapezoidal (TR), and Gear-Shichman (GS) integration formulas as summarized in Table 5.1 are substituted into (5.68) for $V_R'(t_n)$ and $V_R(t_n+h)$ is solved for, the following recursive relationships are obtained:

$$
V_R(t_n+h) = \begin{cases}
(1-\dfrac{h}{\tau})V_R(t_n) & (FE) \\[2em]
(\dfrac{1}{1+h/\tau})V_R(t_n) & (BE) \\[2em]
(1-h/2\tau)/(1+h/2\tau)V_R(t_n) & (TR) \\[2em]
[\dfrac{4}{3}V_R(t_n)-\dfrac{1}{3}V_R(t_n-h)]/(1+2h/3\tau) & (GS)
\end{cases} \tag{5.70}
$$

For the assumed step voltage input, the initial value of $V_R(0) = E_0$ while at time $t = h$, the exact solution is $V_R(h) = E_0 e^{-h/\tau}$.

A comparison of the relative accuracy of the above approximations to the exact solution is most easily made at time $t = 2h$ where for all formulas, the exact solution is assumed known and available at $t = 0$ and $t = h$. This last assumption is necessary because the Gear-Shichman formula requires values at two previous time points. The comparison results are given in Table 5.3 for two choices of h (i.e., $h = 0.1\tau$ and $h = \tau$). It can easily be seen that the Trapezoidal and Gear-Shichman formulas, which are second-order, are more accurate than the Forward and Backward Euler formulas. Further, the loss in accuracy as the stepsize h is increased is also evident. This observed error, referred to as truncation error, is inherent in the methods as will now be shown.

	Exact ($t=h$)	Exact ($t=2h$)	FE ($t=2h$)	BE ($t=2h$)	TR ($t=2h$)	GS ($t=2h$)
$h = 0.1\tau$	0.90483	0.81872	0.81435	0.82256	0.81865	0.81859
$h = \tau$	0.36788	0.13534	0.0	0.18344	0.12263	0.09430

TABLE 5.3 - Comparison of Numerical Integration Formulas ($E_0 \equiv 1.0$)

For the moment, let X_{n+1} be expressed exactly in a Taylor series at $t = (n+1)h$ as

$$X_{n+1} = X_n + h X_n' + \frac{h^2}{2} X_n'' + \frac{h^3}{6} X_n''' + O(h^4) \tag{5.71}$$

where a constant timestep, h, is assumed and $O(h^4)$ represents fourth and higher order terms. Then

$$X_{n+1}' = X_n' + h X_n'' + \frac{h^2}{2} X_n''' + O(h^3) \tag{5.72}$$

If X_{n+1}' from (5.72) is used in the Trapezoidal formula, (5.15), the approximated value of $X_{n+1}(= \hat{X}_{n+1})$ can be expressed as

$$\hat{X}_{n+1} = X_n + \frac{h}{2}(X_n' + X_{n+1}') \tag{5.73}$$

$$= X_n + \frac{h}{2} X_n' + \frac{h}{2} X_n' + \frac{h^2}{2} X_n'' + \frac{h^3}{4} X_n''' + O(h^4) \tag{5.74}$$

$$= X_n + h X_n' + \frac{h^2}{2} X_n'' + \frac{h^3}{4} X_n''' + O(h^4) \tag{5.75}$$

The error introduced at $t = (n+1)h$, ε_{n+1} is given by the difference between the exact value of X_{n+1} from (5.71) and the computed value $\hat{X}_{n+1}$ from (5.75) such that

$$\varepsilon_{n+1} = X_{n+1} - \hat{X}_{n+1} \tag{5.76}$$

$$= \frac{h^3}{6} X_n''' - \frac{h^3}{4} X_n''' + O(h^4) \tag{5.77}$$

$$= -\frac{h^3}{12} X_n''' + O(h^4) \tag{5.78}$$

$$= -\frac{h^3}{12} X_\xi''' \tag{5.79}$$

where $X_\xi''' = X'''(\xi)$ for $n h \le \xi \le (n+1)h$. The resulting error (5.78) or (5.79) arises because the Trapezoidal formula represents an improper truncation of an exact Taylor series expansion. Hence, the term truncation error. Clearly, this error which arose from the use of the Trapezoidal approximation is inherent in the method.

A similar analysis of the Gear-Shichman formula results in truncation error terms of the form

$$\varepsilon_{n+1} = \frac{2}{9} h^3 X_\xi''' \tag{5.80}$$

for a constant timestep h and

$$\varepsilon_{n+1} = \frac{h_1^2 (h_1+h_2)^2}{6(2h_1+h_2)} X_\xi^{'''} \tag{5.81}$$

when different timesteps are used. A comparison of (5.79) with (5.80) indicates that the Trapezoidal formula is characterized by a slightly smaller error constant than the Gear-Shichman formula--a fact born out by inspection of the results in Table 5.3. In fact, it has been demonstrated that the Trapezoidal formula has the smallest truncation error constant of any second order method.

In general, an n th order method will have an associated error term proportional to an $n+1$st derivative evaluated at where $t_n \leq \xi \leq t_n +h$. There are two problems associated with the evaluation of such error terms. The first is being able to accurately evaluate a derivative and the second is knowing where in the interval to evaluate the derivative. As an illustration of the second problem, again consider the example of Figure 5.1 where the Forward and Backward Euler methods are shown. Both are first-order methods and thus characterized by errors proportional to the second derivative. Yet for the timestep illustrated the error for the Forward Euler method is probably more closely approximated if the second derivative is evaluated at t_{n+1}. Admittedly, the example cited is an exaggeration but it does serve to illustrate the problem.

Several techniques are available for approximating a derivative evaluation. One of the most widely used is the divided difference technique. It can be shown that the k th derivative of $X(t)$ can be approximated as

$$\frac{d^k X(\xi)}{dt^k} = k! DD_k \tag{5.82}$$

where DD_k is the k th divided difference. The first divided difference is given by

$$DD_1 = \frac{X_{n+1} - X_n}{h_1} \tag{5.83}$$

$$= \frac{1}{1!} \left[\frac{X_{n+1}}{h_1} + \frac{X_n}{-h_1} \right] \tag{5.84}$$

while the second divided difference can be written as

$$DD_2 = \frac{DD_1(t_{n+1}) - DD_1(t_n)}{h_1 + h_2} \tag{5.85}$$

$$= \frac{1}{2!} \left[\frac{X_{n+1}}{h_1(h_1+h_2)} + \frac{X_n}{(-h_1)(h_2)} + \frac{X_{n-1}}{(-h_1-h_2)(-h_2)} \right] \tag{5.86}$$

In general, the k th divided difference is given by

$$DD_k = \frac{DD_{k-1}(t_{n+1}) - DD_{k-1}(t_n)}{\sum\limits_{i=1}^{k} h_i}$$

(5.87)

$$= \frac{1}{k!} \sum_{i=0}^{k} \frac{X_{n+1-i}}{\prod\limits_{\substack{j=0 \\ j \neq i}}} (t_{n+1-i} - t_{n+1-j})$$

(5.88)

Note that in the above divided difference expressions, the first form defines divided differences in terms of lower order divided differences, while the second form defines divided differences in terms of X and t directly. In either case, a kth order divided difference approximation requires $k+1$ function values or in other words, the retention of k previous solutions. That is, for a second-order method such as Trapezoidal, the solutions X_n, X_{n-1}, and X_{n-2} must be retained for use at t_{n-1} and X_{n-2} must be retained even though they are not used in the integration formula for the evaluation of X_{n+1}.

An alternative procedure is based on simple curve fitting techniques. In effect it makes use of three function values (1 present and 2 previous) plus a present value of the first derivative. By way of illustration, recall that the Gear-Shichman formula for a nonlinear capacitor may be written with an error term included as

$$q_{n+1} = \frac{2}{3} h \, \dot{q}_{n+1} + \frac{4}{3} q_n - \frac{1}{3} q_{n-1} - \frac{2}{9} h^3 q_\xi^{'''}$$

(5.89)

$$= \frac{2}{3} h \, I_{n+1} + \frac{4}{3} q_n - \frac{1}{3} q_{n-1} - \frac{2}{9} h^3 q_\xi^{'''}$$

(5.90)

or in terms of current, I_{n+1}, as

$$I_{n+1} = \frac{3}{2h} q_{n+1} - \frac{2}{h} q_n + \frac{1}{2h} q_{n-1} + \frac{1}{3} h^2 q_\xi^{'''}$$

(5.91)

$$= \hat{I}_{n+1} + \hat{\xi}_{n+1}$$

(5.92)

(It is recognized that what is wanted is $\hat{\xi}_{n+1} \ll \hat{I}_{n+1}$.) Let q be approximated by a third-order polynomial as

$$q = a_3 t^3 + a_2 t^2 + a_1 t + a_0$$

(5.93)

such that

$$q' = 3a_3 t^2 + 2a_2 t + a_1$$

(5.94)

$$q'' = 6a_3 t + 2a_2$$

(5.95)

$$q''' = 6a_3$$

(5.96)

Then at the three time points t_{n-1}, t_n, and t_{n+1}, the following four simultaneous equations may be written

$$q_{n-1} = a_0 + a_1 t_{n-1} + a_2 t_{n-1}^2 + a_3 t_{n-1}^3$$

$$q_n = a_0 + a_1 t_n + a_2 t_n^2 + a_3 t_n^3$$

$$q_{n+1} = a_0 + a_1 t_{n+1} + a_2 t_{n+1}^2 + a_3 t_{n+1}^3 \tag{5.97}$$

$$\dot{q}_{n+1} = a_1 + 2a_2 t_{n+1} + 3a_3 t_{n+1}^2$$

from which a_3 can be evaluated where it is assumed

$$\dot{q}_{n+1} = \hat{I}_{n+1} \tag{5.98}$$

The system of equations (5.79) can be further simplified by shifting the function values on the time axis such that

$$t_{n-1} = 0$$

$$t_n = h_2 \tag{5.99}$$

$$t_{n+1} = h_1 + h_2$$

in which case (5.97) reduces trivially to

$$\begin{bmatrix} \dot{q}_{n+1} \\ q_{n+1} - q_{n-1} \\ q_n - q_{n-1} \end{bmatrix} = \begin{bmatrix} 1 & 2(h_1+h_2) & 3(h_1+h_2)^2 \\ h_1+h_2 & (h_1+h_2)^2 & (h_1+h_2^3) \\ h_2 & h_2^2 & h_2^3 \end{bmatrix} \begin{bmatrix} a_1 \\ a_2 \\ a_3 \end{bmatrix} \tag{5.100}$$

For this approach, it can be seen that all previous solutions retained for computation of the truncation error are used in the integration formula itself for Gear-Shichman while only one, X_{n-1}, is unused in the Trapezoidal formula.

Now that much has been said about truncation error and approximate methods for its computation, there still remains a significant question regarding its importance. Unfortunately, no simple answer exists. The relative effect on any given node voltage, branch voltage or branch current in one portion of the circuit to an induced truncation error at a reactive element in another part of a circuit depends critically on the nature of the circuit. For example, given a unilateral gain stage with a reactive load, a very large amount of truncation error at the load will have no effect on the input circuitry. If the response at the input is all that's desired, then the truncation error is of no importance. On the other hand, if the situation is reversed, then a small error at the input may have disastrous consequences. Clearly, what is required is a means of determining the sensitivity of those circuit variables considered to be of importance to the truncation error terms associated with reactive elements. In a later chapter on sensitivity and adjoint networks it will be shown that effective methods do exist for calculating dc and small- signal sensitivities of any basic circuit variables to perturbations in element values or other circuit variables. Unfortunately, such sensitivity calculations in the large- signal transient or time domain are much more costly and hence of limited effectiveness. Nonetheless, assume such sensitivity calculations were cheap and effective. Would it then be possible to assess accurately the importance of truncation error to output variables? In general the answer is still no. The following example illustrates the problem: Consider a differential amplifier in which one side is over-driven thus dominating the output while the other input is reactively loaded and driven in such a way that its currents and voltages are significantly in error due to truncation error. A sensitivity analysis would indicate the output to be relatively unaffected by the truncation error because of the over driven condition on the other input. If now the overdriven condition is removed before the induced error at the other input has settled out, a significant error at the output may result which no amount of timestep reduction will alleviate. A similar situation can arise in digital circuits, say for a dual input NAND gate.

The predicament just described above leads to the conclusion that the only safe course is to insist that truncation error terms be maintained at a negligible level on an element by element basis so that they are given no chance to propagate. In general, this result is achieved by controlling the size of the timestep h, reducing it as necessary to maintain a small relative error and expanding it whenever possible to speed the total simulation. This subject of timestep control will be considered in more detail following an introduction to stability concepts.

5.5 Stability of Integration Methods

In contrast to truncation error which was shown to be a localized property (local to the present time point and timestep), stability is a global property related to the growth or decay of errors introduced at each time point and propagated to successive time points. It is important to realize that in general stability depends *both* on the method *and* on the problem.

If the example of the previous section is again referred to, intuitively one would expect that for any applied method, stability would be characterized by the computed solution approaching the exact solution as time proceeds to infinity. The first three methods of (5.70) can be used to qualitatively illustrate the constraints derivable from this intuitive viewpoint. At $t_n = nh$,

$$V_R(nh) = \begin{cases} E_0(1 - \dfrac{h}{\tau})^n & (FE) \\[2mm] E_0/(1 + \dfrac{h}{\tau})^n & (BE) \\[2mm] E_0(1 - \dfrac{h}{2\tau})^n / (1 + \dfrac{h}{2\tau})^n & (TR) \end{cases} \qquad (5.101)$$

By inspection, it can be seen that if $h > 2\tau$ for the Forward Euler method, an unbounded (and incorrect) solution results. In contrast, the Backward Euler method tends to the correct result of zero regardless of the size of h. (It is understood that the timestep h is positive.) This comparison loosely supports the conjecture that implicit methods are more stable than explicit methods.

For the Trapezoidal formula, a comparison with the Backward Euler formula indicates that for large h, the ratio $(1 - h/2\tau)^n / (1 + h/2\tau)^n$ remains less than 1 such that the solution does tend toward zero. However, for $h > 2\tau$, it does so in a oscillatory manner which may not be very satisfying. Further, it may converge slowly. This comparison supports the conjecture that higher-order methods, even though implicit, are less stable than lower order methods.

While the example above loosely illustrates the idea of stability, the definition of stability needs to be further refined. Formal definitions for both absolute and relative stability exist [B9], [H14] but are usually phrased in terms of errors not being allowed to propagate unbounded rather than in terms of convergence to the exact solution. Further, as will be brought out shortly, detailed stability analysis usually relies on the use of difference equations which are non-linear for non-linear problems.

Since the general stability analysis problem is difficult and unwieldy, an approach usually adopted is to compare the stability of different methods for a single test equation

$$\dot{X} = \lambda X \qquad\qquad (5.102)$$

(Note from (5.68) and (5.69) that the example which has been considered is of this form where the eigenvalue $\lambda = -1/\tau$.) Further, for the above test equation several types of stability have been defined. Dahlquist [H2] has defined an integration algorithm to be *A-stable* if it results in a stable difference equation approximation to a stable differential equation. Said another way, a method is A-stable if all numerical approximations tend to zero as $n \to \infty$ when applied to (5.102) for fixed positive h and an eigenvalue, λ, in the left-half plane. This definition is equivalent to the requirement that the region of stability of the difference equation approximation include the open left-half of the $h\lambda$ plane (or right-half of the h/τ plane) and implies that for a stable differential equation an A-stable method will converge to the exact solution as time goes to infinity regardless of the stepsize or the accuracy at intermediate steps.

Before actually comparing the four methods in (5.70) with respect to A-stability, it is useful to introduce a systematic approach to derive stability constraints. For the Trapezoidal formulation, (5.70) may be rewritten in the more general form of a recursion relation as

$$(1 - h\lambda/2)V_n - (1 + h\lambda/2)V_{n-1} = 0 \qquad\qquad (5.103)$$

where λ has been substituted for $-1/\tau$. This equation is a linear homogeneous difference equation with constant coefficients. As such, every solution can be written in the form

$$V_n = C_1 X_1^n + \cdots + C_k X_k^n \tag{5.104}$$

where $X_1, X_2, \cdots, X_k$ are k distinct roots of the characteristic polynomial

$$p_1(X) = (1 - h\lambda/2)X - (1 + h\lambda/2) = 0 \tag{5.105}$$

(In this case, $k = 1$.) Thus,

$$X_1 = \frac{(1 + h\lambda/2)}{(1 - h\lambda/2)} \tag{5.106}$$

such that

$$V_n = C_1 X_1^n = C_1 \left[\frac{(1 + h\lambda/2)}{(1 - h\lambda/2)} \right]^n \tag{5.107}$$

But

$$V_0 = C_1 X_1^0 = C_1 = E_0 \tag{5.108}$$

such that

$$V_n = \left[\frac{(1 + h\lambda/2)}{(1 - h\lambda/2)} \right]^n E_0 \tag{5.109}$$

which is consistent with (5.101).

For the Forward Euler, Backward Euler and Trapezoidal formulations it is easily seen that in each case a first-order difference equation resulting in a single root will be obtained. If (5.102) is assumed to have been generalized beyond the original RC circuit example such that the eigenvalue λ is allowed to take on values anywhere in the complex plane, then for these three formulations, the derivable constraints lead to the following conditions for stability:

$$\left| 1 + h\lambda \right| < 1 \quad \text{or} \quad \left| 1 - h/\tau \right| < 1 \qquad (FE)$$

$$1 < \left| 1 - h\lambda \right| \quad \text{or} \quad 1 < \left| 1 + h/\tau \right| \qquad (BE) \tag{5.110}$$

$$\left| \frac{(1 + h\lambda/2)}{(1 - h\lambda/2)} \right| < 1 \text{ or } \left| \frac{(1 - h/2\tau)}{(1 + h/2\tau)} \right| < 1 \quad (TR)$$

For Forward Euler, (5.110) represents the region *inside* a unit circle at -1 in the $h\lambda$-plane or at $+1$ in the h/τ-plane. For a real eigenvalue λ this region reduces to an interval on the $h\lambda$-axis given by $-2 < h\lambda < 0$ or equivalently $0 < h/\tau < 2$ which is consistent with the earlier observation that for asymptotic convergence, $h < 2\tau$. For Backward Euler, (5.110) represents the region *outside* a unit circle at $+1$ in the $h\lambda$-plane or at -1 in the h/τ-plane. Again, for a real eigenvalue λ this region reduces to the intervals on the $h\lambda$-axis given by $h\lambda < 0$ and $2 < h\lambda$ or equivalently $0 < h/\tau$ and $h/\tau < -2$ where the first interval is again consistent with the earlier observation. For the Trapezoidal formulation, (5.110) can be shown to represent the open left-half of the $h\lambda$-plane or the open right-half of the h/τ-plane.

Next, consider the Gear-Shichman formulation of (5.70) which may be written in the form of a recursion relation as

$$(1 + \frac{2h}{3\tau}) V_n - \frac{4}{3} V_{n-1} + \frac{1}{3} V_{n-2} = 0 \qquad (5.111)$$

This linear homogeneous second-order difference equation has an associated characteristic polynomial given by

$$p_2(X) = (1 + \frac{2h}{3\tau}) X^2 - \frac{4}{3} X + \frac{1}{3} \qquad (5.112)$$

A simplification results if (5.112) is transformed into the polynomial

$$p_2(Y) = \frac{1}{3} Y^2 - \frac{4}{3} Y + (1 + \frac{2h}{3\tau}) \qquad (5.113)$$

where $Y = 1/X$. The roots of $p_2(Y)$ are given by

$$Y_{1,2} = 2 \pm \sqrt{1 - \frac{2h}{\tau}} \qquad (5.114)$$

such that

$$X_{1,2} = \frac{1}{2 \pm \sqrt{1 - \frac{2h}{\tau}}} \qquad (5.115)$$

It can be shown that where there are multiple, distinct roots, one root, referred to as the principle root, most closely approximates the true solution and dominates the remaining roots as a condition of stability. If the dominant root is denoted arbitrarily as X_1, then it can be shown for the coefficients in (5.104) that $C_1 \gg C_2, C_3$, etc.[H20] Finally, it can also be shown that based upon the consistency condition the principle root can be identified by the constraint

$$X_1 \rightarrow 1 \quad \text{for} \quad h \rightarrow 0 \qquad (5.116)$$

Thus

$$X_1 = \frac{1}{2 - \sqrt{1 - \frac{2h}{\tau}}} \quad \text{and} \quad X_2 = \frac{1}{2 + \sqrt{1 - \frac{2h}{\tau}}} \qquad (5.117)$$

For the Gear-Shichman formula to result in a stable solution it is necessary that $|X_1| < 1$ which implies

$$\left| 2 - \sqrt{1 - \frac{2h}{\tau}} \right| > 1 \qquad (5.118)$$

This condition doesn't have an obvious geometrical interpretation but can in fact be shown [H14] to be approximately a circle of radius 2 centered at +2 on the $h\lambda$-plane. Again, if consideration is given to real eigenvalues λ, (5.118) can be re-stated as

$$2 - \sqrt{1 - \frac{2h}{\tau}} > 1 \quad \text{and} \quad -2 + \sqrt{1 - \frac{2h}{\tau}} > 1 \qquad (5.119)$$

These two constraints can be reduced to show that the Gear-Shichman formula is stable on the intervals $h\lambda < 0$ and $4 < h\lambda$ on the $h\lambda$-axis, which is consistent with the region described above, or $0 < h/\tau$ and $h/\tau < -4$ on the h/τ-axis.

For the four methods examined in detail and for real eigenvalues, the conditions for stability just derived are summarized in Table 5.4.

METHOD	STABILITY REGION
Forward Euler	$0 < \dfrac{h}{\tau} < 2$
Backward Euler	$\dfrac{h}{\tau} < -2 \quad \text{and} \quad 0 < \dfrac{h}{\tau}$
Trapezoidal	$0 < \dfrac{h}{\tau}$
Gear-Shichman	$\dfrac{h}{\tau} < -4 \quad \text{and} \quad 0 < \dfrac{h}{\tau}$

TABLE 5.4 - Stability Regions for Common Numerical Integration Formulas

Overall it can be observed that the Backward Euler formulation is characterized by the largest region over which it is stable followed by the Gear-Shichman, Trapezoidal, and Forward Euler formulations, respectively. Lest too much weight be placed on the above results, one should realize that they apply only for the special test equation (5.102) or equivalently for a single time-constant linear circuit. Numerical integration formulas do not possess regions of stability independent of the problem to which they are applied. Furthermore, while the approach used above of transforming a recursive relationship into a difference formula in order to find a general solution remains valid for nonlinear problems, as previously pointed out, the resulting difference equation will also be nonlinear, without a simple closed form solution. Finally, for the case of multiple time constant (eigenvalue) circuits it can be shown that whatever constraint is placed upon the stepsize h in order to maintain stability is determined by the largest eigenvalue, (smallest time constant). By contrast, the response of interest is very often determined by the smallest eigenvalue (largest time constant). These two facts lead to the conflicting requirements of a long simulation with short timesteps (i.e., many analysis points). Again, a compromise between higher-order and consequently more accurate formulas with smaller regions of stability which demand small timesteps and low order methods of less accuracy but large regions of stability which permit large timesteps is required.

Gear [H5,H14] recognized that Dahlquist's definition of A-stability excludes all methods of order greater than two and many of order two. His definition of "stiff stability" illustrated in Figure 5.3 represents a practical compromise. The rationale for this definition is based on the following: The change in a single timestep due to an eigenvalue λ is $e^{h\lambda}$. If $h\lambda = \sigma + j\omega$, then the change in magnitude is e^{σ}. Thus, in Region 1 the component is reduced by at least e^{σ} in a single step and therefore likely remain small. In this case, accuracy is of little importance such that it it possible to ignore all components in Region 1 for some α. The only requirement is absolute stability (i.e. that no component blow up). In Region 2, an accurate, stable method is required for the following reasons: For $\alpha < \sigma < 0$, components decay slowly such that to prevent accumulation of error solutions must be computed accurately. For $0 < \sigma < \phi$, components increase rapidly such that small steps must be taken to accurately track these changes. Finally, for $|\omega| > \theta$, at least $\theta/2\pi$ periods of oscillation must occur such that again small steps must be taken to model these oscillations. In either of the latter two cases the accuracy concomitant with a small timestep precludes accumulation of error. Gear has shown that backward difference formulas of up to order 6 exist which are stiffly stable. Their regions of stability are illustrated in Figure 5.4. Practically speaking, stiffly-stable methods are characterized by the properties that they are (1) stable as the timestep approaches infinity and (2) stable when λ is real and negative.

An alternative plausibility argument [46] in support of stiff stability proceeds as follows: If the contribution of the four poles of Figure 5.5 to a transient solution are considered, it is clear that λ_4 places the most severe demands on the numerical process. It should be recognized that the previously considered h,λ-plane is nothing more than a scaled λ-plane where h is the scale factor. As a consequence, it can be postulated that a method need not be stable in the vicinity of λ_4 since small timesteps will be required for accuracy anyway. On the other hand, if λ_1 and λ_2 and/or λ_3 and then a large timestep to accommodate λ_1. But these requirements precisely agree with the previous definition of stiff stability.

5.6 Automatic Timestep Control

It has been previously pointed out that in any transient simulation, the computation time will be proportional to the number of time increments into which the simulation must be divided. Typically, the simulation time is a multiple of the largest time constant (smallest eigenvalue) associated with the linearized circuit. Conversely, the limitation on the timestep h is determined by the smallest time constant, (largest eigenvalue) of the linearized circuit. Thus arise the conflicting requirements of a long simulation time with a small timestep, leading to a large computation time. Systems such as this have been classified as stiff systems.

Numerous techniques have been proposed for monitoring the "activity" within a system of differential equations in order to determine the maximum allowable stepsize consistent with stability and truncation error. It has previously been shown that bounds on the timestep h may be due to the method (truncation error) or the problem (spread and location of eigenvalues). With this realization in mind, several authors have proposed that both the order of the method and the timestep be allowed to vary in order to achieve the largest possible timestep consistent with accuracy. The

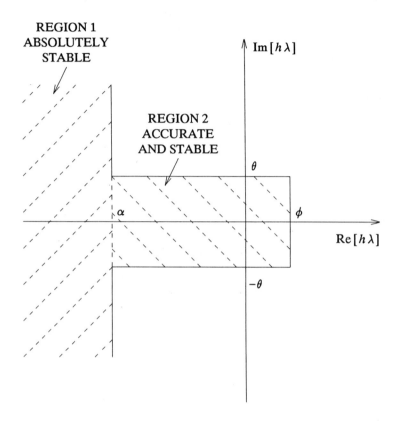

Figure 5.3 - Illustration of requirements for stiff stability

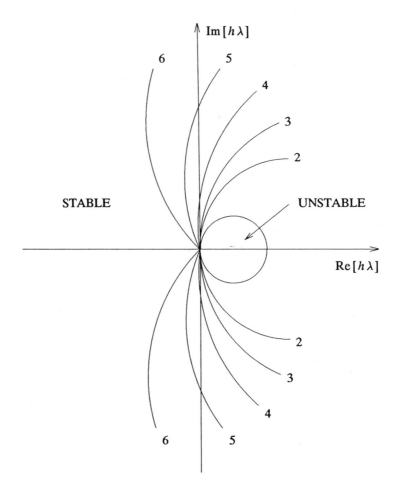

Figure 5.4 - Regions of stability for stiffly stable backward difference through sixth order

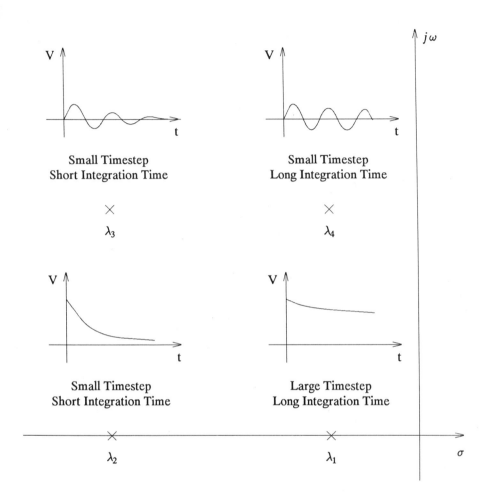

Figure 5.5 - Effect of pole locations on stability requirements

rationale for this approach is illustrated in Figure 5.6. Recall that for a kth order method, the truncation error incurred is given by

$$\varepsilon_k \propto h^{k+1} X_\xi^{(k+1)} \tag{5.120}$$

For sufficiently small values of h, ε_k decreases as the order k increases because higher-order derivatives tend to zero. On the other hand, as the timestep h is increased, a point is reached where $\varepsilon_k > \varepsilon_{k-1}$ because $h^{k+1} >> h^k$ or $h^{k+1} X_\xi^{(k+1)} > h^k X_\xi^{(k)}$. Thus in Figure 5.6, if the allowable error per timestep is ε_1, the first-order method provides the largest timestep. If the allowable error per timestep is ε_2, the method of order two maximizes h. Finally, if a tight error requirement is to be met such as ε_3, the third order method should be used. The variable-order, variable-step approach is based therefore on the computation of local truncation error at each time point for a method of order k *and* the methods of order k-1 and order k+1 as well. For the next time point, the order of the method is chosen so as to maximize h while maintaining required accuracy. Nagel [C13] and others have reported experimental results indicating that for integrated circuits the order most often selected is two. Consequently, SLIC, SINC, SPICE and SPICE-2 all use fixed, second-order methods.

Given that a second-order (or any other fixed-order) method is decided upon, it is still possible and desirable to vary the timestep taking small timesteps when network variables are undergoing transitions in order to preserve accuracy and large timesteps when there is little activity. Here again, truncation error can and has been used to monitor circuit activity.

The desirability of using such an approach to bound and control accumulated error while simultaneously providing a guide to the selection of the largest possible timestep is somewhat offset by the difficulty of computing a realistic value for $X_\xi^{(k+1)}$. As previously indicated, some form of divided difference or least square curve fitting approximation can be used. However, such estimates usually tend to be overly pessimistic and usually result in restrictively small values of timestep h.

An alternate approach to timestep control or selection is based on the recognition that the conditions of slow voltage and current variations within a circuit which facilitate a large timestep are precisely the same conditions under which a Newton-Raphson iteration procedure converges quickly to the solution of the system of nonlinear algebraic equations. Hence, timestep selection is coupled to iteration count at each new time point in a scheme similar to the following: If a solution is reached in five or fewer Newton iterations, the timestep is doubled. If ten or more iterations are required, the timestep is halved. Otherwise the timestep remains the same.

It is interesting to compare the Trapezoidal and Gear-Shichman formulas with regard to the effect of halving or doubling the timestep. Assume that h_0 has been the existing timestep and that h is the new timestep. Then

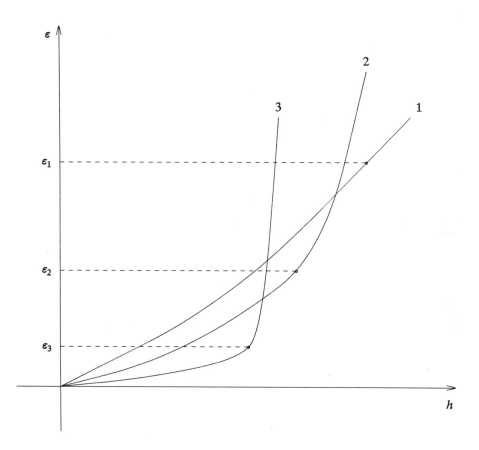

Figure 5.6 - Error as a function of timestep and order

$$\varepsilon_n \propto \begin{cases} \dfrac{1}{12} h_0^3 & (TR) \\[4mm] \dfrac{2}{9} h_0^3 & (GS) \end{cases} \tag{5.121}$$

If the timestep is halved such that $h = h_0/2$

$$\varepsilon_{n+1} \propto \begin{cases} \dfrac{1}{96} h^3 & (TR) \\[4mm] \dfrac{3}{64} H_0^3 & (GS) \end{cases} \tag{5.122}$$

where the latter Gear-Shichman expression is based on the variable timestep truncation error prediction formula of (5.81). The resulting decrease in error then is

$$\varepsilon_{n+1}/\varepsilon_n \propto \begin{cases} 1/8 & (TR) \\[4mm] ^-1/5 & (GS) \end{cases} \tag{5.123}$$

Similarly if the timestep is doubled where $h = 2h_0$

$$\varepsilon_{n+1} \propto \begin{cases} \dfrac{2}{3} h_0^3 & (TR) \\[4mm] ^-1.2 h_0^3 & (GS) \end{cases} \tag{5.124}$$

such that

$$\varepsilon_{n+1}/\varepsilon_n \propto \begin{cases} 8 & (TR) \\[4mm] ^-5.4 & (GS) \end{cases} \tag{5.125}$$

Naturally, the above results tend to ignore the fact that X_ξ''' will vary as the timestep is halved or doubled; however, the results do lead to two interesting and opposing views. The first is that because halving or doubling a timestep results in nearly an order of magnitude change in error for the Trapezoidal method, the method can respond rapidly to changing conditions within the circuit simulation. Hence, it is a better method than the Gear-Shichman formulation. A counter argument is that the Trapezoidal method reacts too violently and thus results in an erratic simulation in comparison to the Gear-Shichman method. Clearly, both arguments are based only on conjecture.

To conclude this section on timestep selection and control, several additional techniques, applicable to either truncation error or iteration count timestep control, for speeding up transient analyses (reducing the number of Newton iterations) are considered. The first such technique is based on the identification of those time points in the simulation at which piece-wise linear and/or pulsed input sources have breakpoints; that is, points of discontinuity at the beginning or end of a switching transition. The timestep is always reduced as such breakpoints are encountered to insure that the breakpoint itself is used as a time point. In addition, the first timestep taken beyond the breakpoint is arbitrarily set at 1/10 the impending rise or fall time of the pulse. This technique assures accurate location of switching transitions and the selection of an initial timestep commensurate with the speed of the transition.

A second technique used to reduce the total number of Newton iterations is based on the prediction of the solution at time t_{n+1} based on previous solution values at time t_n, t_{n-1}, etc. in order to provide a better starting point for the Newton-Raphson iteration algorithm. In essence, the first Newton iterate is obtained via extrapolation from previous values as follows: A polynomial of order k which interpolates to $k+1$ known solution points $(X_n, X_{n-1}, ..., X_{n-k})$ is evaluated at t_{n+1} to obtain an initial prediction of X_{n+1}. If the polynomial is written as

$$p_k(t) = a_k t^k + a_{k-1} t^{k-1} + \cdots + a_1 t_k + a_0 \tag{5.126}$$

and t is taken as $t_n, t_{n-1}, ..., t_{n-k}$, the system of linear equations

$$a_k t_n^k + a_{k-1} t_n^{k-1} + \cdots + a_1 t_n + a_0 = X_n$$

$$a_k t_{n-1}^k + a_{k-1} t_{n-1}^{k-1} + a_1 t_{n-1} + a_0 = X_{n-1} \tag{5.127}$$

$$\cdots$$

$$a_k t_{n-k}^k + a_{k-1} t_{n-k}^{k-1} + \cdots + a_1 t_{n-k} + a_0 = X_{n-k}$$

is obtained. The initial guess of X_{n+1} is then given by

$$X_{n+1}^{(0)} = P_k(t_{n+1}) \tag{5.128}$$

An alternative procedure which will lead to the same initial estimate or guess of X_{n+1} is to generate an explicit integration formula for which all b_j in (5.46) equal zero. That is, find a set of coefficients c_i such that

$$X_{n+1} = \sum_{i=0}^{k} c_i X_{n-i} \tag{5.129}$$

By analogy with (5.61) the c_i are obtained as the solution to the system of equations

$$1 = c_0 + c_1 + \quad + c_k$$

$$0 = (-h_1) c_0 + (-h_1 \cdot h_2) c_1 + \cdots + (-h_1 \cdot h_2 \cdots - h_{k+1}) c_k \tag{5.130}$$

$$\cdots$$

$$0 = (-h_1)^k c_0 + (-h_1 - h_2)^k c_1 + \cdots + (-h_1 \cdot h_2 \cdots -h_{k+1})^k c_k$$

where $h_1 = t_{n+1} - t_n, h_2 = t_n - t_{n-1}$, etc. By way of example, Shichman has derived the following predictor formula for use with the Gear-Shichman method:

$$
\begin{aligned}
X_{n+1}^{(0)} = {}& \frac{h_1(h_1 + h_2)}{h_3(h_2 + h_3)} X_{n-2} \\[2mm]
& - \frac{h_1(h_1 + h_2 + h_3)}{h_2 h_3} X_{n-1} \\[2mm]
& + \frac{(h_1 + h_2)(h_1 + h_2 + h_3)}{h_2(h_2 + h_3)} X_n
\end{aligned}
\tag{5.131}
$$

Note that this strategy of using a predictor equation to obtain an initial guess at X_{n+1} is *not* the same as what is traditionally referred to as a predictor-corrector integration method. This strategy instead is based on the use of Newton-Raphson to iteratively solve the implicit system of equations for X_{n+1}.

A third technique for speeding up transient analyses and reducing the number of Newton iterations is to ignore the print interval specified by the user and to take whatever timestep is possible. The solutions at each time point are stored on a temporary disk file for later retrieval. At the end of the simulation the results are read back from disc for printing and plotting. At this point, an interpolation algorithm is used to obtain values at the evenly spaced intervals requested by the user. A particular effective approach is to use the same technique as was used to derive the previous predictor formula (5.129). In the second-order case, this procedure amounts to solving (5.131) for X_n and letting X_{n+1}, X_{n-1} and X_{n-2} be the computed points while X_n becomes the interpolated point. Alternatively, (5.131) itself can be used where X_n, X_{n-1} and X_{n-2} are the computed points and t_{n+1} lies between t_n and t_{n-1} such that $h_1 = t_{n+1} - t_n < 0$.

At this point, it is worthwhile to outline the steps required in a transient analysis in order to point out several additional speed-up techniques. The basic steps are shown in Figure 5.7. Note that two iteration loops are now present--an inner loop for the required implicit Newton-Raphson iteration and an outer loop for selection and advancement of time via an integration formula. It has been observed that of the total time spent in performing a Newton iteration, 80% is spent linearizing and evaluating nonlinear elements and 20% in solving linear equations. As a consequence it is most desirable to remove any unnecessary computation from the inner Newton-Raphson loop in Figure 5.7. An approach which has been used successfully is to treat nonlinear reactive elements as linear reactive elements at time t_{n+1} whose values are those computed from X_n at time t_n. This procedure introduces small error so long as the circuit operation varies little between times t_n and t_{n+1}, a requirement which can be met by effective timestep control. The result is that the second-step in the inner Newton-Raphson loop is moved outside the iteration.

It seems reasonable to expect that if the above procedure works, that it can be extended to other, secondary effects such as the variation of Beta with current, basewidth modulation and high-level injection effects. If this line of reasoning is carried to its extreme, the conclusion is that the entire Jacobian (admittance/conductance matrix) could be evaluated once at time t_n, factored into L and U for the first iteration at time t_{n+1} and retained. Each iteration at t_{n+1} after the first would require a re-evaluation of the RHS current vector only and then a forward and backward substitution. The fact that the current vector must change can be seen by recalling that for a nonlinear element, the Norton equivalent current source is given by

$$I_N^{(n+1)} = I^{(n)} - G^{(0)} V^{(n)} \qquad (5.132)$$

where in this case $I^{(n)}$ and $V^{(n)}$ are changing with each iteration while $G^{(0)}$ was computed at t_n. It is essential to realize that the potential success of this approach is based on the fact that in the Newton-Raphson iteration algorithm described by

$$V^{(n+1)} = V^{(n)} - \frac{\overline{G}(V^{(n)})}{\overline{G}'(V^{(n)})} \qquad (5.133)$$

it is not essential for $\overline{G}'(V^{(n)})$ to be exact for convergence. That is, $V^{(n+1)} = V^{(n)}$ for $\overline{G}(V^{(n)}) = 0$ regardless of the accuracy of $\overline{G}'(V^{(n)})$, the Jacobian. The modified transient analysis procedure for this approach is outlined in Figure 5.8.

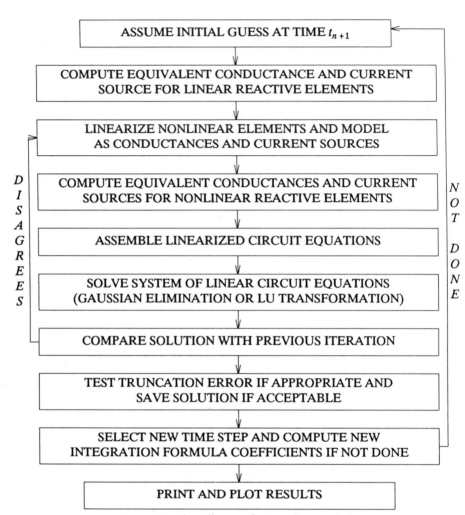

Figure 5.7 - Flow diagram for transient analysis

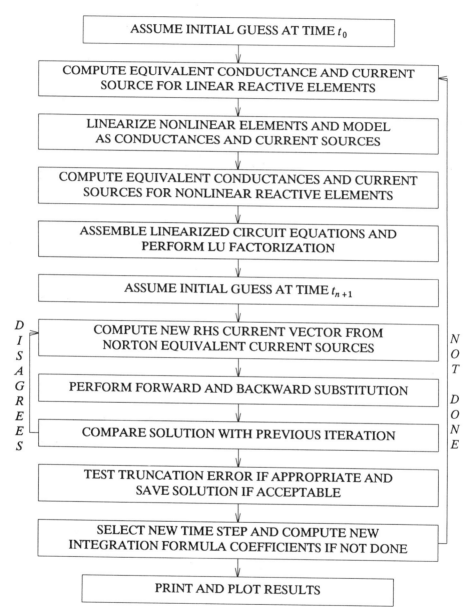

Figure 5.8 - Flow diagram for modified transient analysis

6. ADJOINT NETWORKS AND SENSITIVITY

6.1 Adjoint Networks

The adjoint network concept developed by Director and Rohrer [J8,J10] has proved useful in rapidly and accurately calculating element sensitivities (partial derivatives), noise performance, group delay, etc. Its derivation is usually based on Tellegen's Theorem but can proceed from the use of calculus of variations. The result is that either the partial derivatives of an output quantity with respect to many element values or the transfer functions from many input ports to a single output port can be easily computed.

Given topologically equivalent networks (same directed graph) η and $\hat{\eta}$ where η is the original network and $\hat{\eta}$ is the adjoint network (to be defined below), Tellegen's Theorem implies

$$\sum_B V_B I_B = 0 \qquad \sum_B \psi_B \phi_B = 0 \qquad\qquad (6.1)$$

In (6.1) V_B and I_B are the respective branch voltages and currents in the original network η and ψ_B and ϕ_B are the corresponding branch voltages and currents in the adjoint network $\hat{\eta}$. For all branches including sources the sign convention of Figure 6.1 is observed.

Figure 6.1 - Standard branch sign conventions

Equations (6.1) can be thought of as a statement of conservation of energy or power where sources consume negative power and elements like resistors, diodes etc. consume positive power. More important to the present derivation is the fact that Tellegen's Theorem also implies

$$\sum_B V_B \, \phi_B = 0 \qquad \sum_B \psi_B \, I_B = 0 \tag{6.2}$$

Assume that the original network is perturbed slightly such that $V_B \rightarrow V_B + \delta V_B$ and $I_B \rightarrow I_B + \delta I_B$ then from (6.2)

$$\sum_B (V_B + \delta V_B) \phi_B = 0 \qquad \sum_B \psi_B \, (I_b + \delta I_B) = 0 \tag{6.3}$$

Subtraction of (6.2) from (6.3) results in

$$\sum_B \delta V_B \, \phi_B = 0 \qquad \sum_B \psi_B \, \delta I_B = 0 \tag{6.4}$$

Since (6.4) is true, it is also true that

$$\sum_B \delta V_B \, \phi_B - \psi_B \, \delta I_B = 0 \tag{6.5}$$

Equation (6.5) is the basic starting point from which all individual element sensitivities are derived.

6.2 Element Sensitivity

The sensitivity of an output quantity to a small-change or perturbation in any element value can be derived as follows: For a network with a single source and a single output, (6.5) may be rewritten as

$$
\begin{aligned}
\delta V_S \, \phi_S - \psi_S \, \delta I_S \; &+ \; \delta V_O \, \phi_O - \psi_O \, \delta I_O \\[4pt]
&+ \; \sum_G \delta V_G \, \phi_G - \psi_G \, \delta I_G \\[4pt]
&+ \; \sum_C \delta V_C \, \phi_C - \psi_C \, \delta I_C \\[4pt]
&+ \; \sum_\Gamma \delta V_\Gamma \phi_\Gamma - \psi_\Gamma \delta I_\Gamma \\[4pt]
&+ \; \sum_{G_T} \delta V_I \, \phi_I - \psi_I \, \delta I_I + \delta V_D \, \phi_D - \psi_D \, \delta I_D \\[4pt]
&= 0
\end{aligned}
\tag{6.6}
$$

where the subscript S represents the source branch; the subscript O represents the output branch; the summation over G represents all conductances; C, all capacitances; Γ, all reciprocal inductances; and G_T, all voltage-controlled current sources. For the latter, the subscript I represents the independent or controlling branch and the subscript D represents the dependent or controlled branch. For the moment consider a single conductance with the branch relation

$$I_G = G V_G \tag{6.7}$$

If G is perturbed such that $G \to G + \delta G$, then (6.7) becomes

$$I_G + \delta I_G = (G + \delta G)(V_G + \delta V_G) \tag{6.8}$$

$$= G V_G + G \delta V_G + \delta G V_G + \delta G \delta V_G \tag{6.9}$$

$$\approx G V_G + G \delta V_G + \delta G V_G \tag{6.10}$$

where only first-order terms are retained in (6.10). Subtraction of (6.7) from (6.10) results in

$$\delta I_G = G \delta V_G + \delta G V_G \tag{6.11}$$

If (6.11) is substituted in the summation over conductances in (6.6), the result is

$$\sum_G \delta V_G \phi_G - \psi_G \delta I_G = \sum_G \delta V_G \phi_G - \psi_G G \delta V_G - \psi_G \delta G V_G$$

$$= \sum_G (\phi_G - G \psi_G) \delta V_G - V_G \psi_G \delta G$$

$$= \sum_G (\ 0\ ) \delta V_G - V_G \psi_G \delta G$$

$$= -\sum_G V_G \psi_G \delta G \tag{6.12}$$

The cancellation in the expression leading to (6.12) is based upon letting every conductance branch G in the original network η correspond to a similar conductance branch G in $\hat{\eta}$ such that the branch relation in $\hat{\eta}$ is

$$\phi_G = G \psi_G \tag{6.13}$$

Similarly, it is found for a capacitor with an admittance branch relation

$$I_C = j\omega C V_C \tag{6.14}$$

that

$$\sum_C \delta V_C \phi_C - \psi_C \delta I_C = -\sum_C j\omega V_C \psi_C \delta C \tag{6.15}$$

where in the adjoint network

$$\phi_C = j\omega C \psi_C \tag{6.16}$$

and for an inductor with a branch relation

$$I = \frac{\Gamma}{j\omega} V_\Gamma \tag{6.17}$$

that

$$\sum_\Gamma \delta V_\Gamma \phi_\Gamma - \psi_\Gamma \delta I_\Gamma = -\sum_\Gamma \frac{1}{j\omega} V_\Gamma \psi_\Gamma \delta\Gamma \tag{6.18}$$

where in the adjoint network

$$\phi_\Gamma = \frac{\Gamma}{j\omega}\psi_\Gamma \tag{6.19}$$

Finally, if the branch relation for a voltage-controlled current source or transconductance in the original network η is

$$I_D = G_T V_I \tag{6.20}$$

such that

$$\delta I_D \approx G_T \delta V_I + \delta G_T V_I \tag{6.21}$$

and if the corresponding branch relation in the adjoint network is

$$\phi_I = G_T \psi_D \tag{6.22}$$

then the summation over transconductances in (6.6) becomes

$$\sum_{G_T} \delta V_I \phi_I - \psi_I \delta I_I + \delta V_D \phi_D - \psi_D \delta I_D$$

$$= \sum_{G_T} \delta V_I \phi_I - \psi_I \delta I_I + \delta V_D \phi_D - \psi_D G_T \delta V_I - \psi_D \delta G_T V_I$$

$$= \sum_{G_T} (\phi_I - G_T \psi_D)\delta V_I - \psi_I \delta I_I + \delta V_D \phi_D - V_I \psi_D \delta G_T$$

$$= \sum_{G_T} (\ 0\ )\delta V_I - \psi_I 0 + \delta V_D 0 - V_I \psi_D \delta G_T$$

$$= -\sum_{G_T} V_I \psi_D \delta G_T \tag{6.23}$$

As before, the cancellations leading to (6.23) result by identifying a transconductance within the adjoint network as given by the branch relation (6.22) where it can be seen that the roles of dependent and independent variables have been reversed. Thus for the adjoint network, ϕ_D, which represents the current through an open circuit, is zero while in the original network, I_I represents the current through the controlling open-circuit such that δI_I is zero. Note that as a current source represents a generalized open circuit, both the controlling and controlled branches in both the original and adjoint networks are open circuits. In general, though not brought out here, short circuits in the original network remain short circuits in the adjoint network while open circuits in the original network remain open circuits in the adjoint network such that current-controlled current sources in the original network become voltage-controlled voltage sources in the adjoint network and vice-versa while current-controlled voltage sources remain current-controlled voltage sources. In all cases the roles of controlling and dependent branches are reversed.

With the substitution of (6.12), (6.15), (6.18) and (6.23) in (6.6), (6.6) can be rewritten as

$$\delta V_S \, \phi_S - \psi_S \, \delta I_S + \delta V_O \, \phi_O - \psi_O \, \delta I_O$$

$$= \sum_G V_G \, \psi_G \, \delta G + \sum_C j\omega V_C \, \psi_C \, \delta C + \sum_\Gamma \frac{1}{j\omega} V_\Gamma \, \psi_\Gamma \, \delta\Gamma + \sum_{G_T} V_I \, \psi_D \, \delta G_T \qquad (6.24)$$

In computing the sensitivity (partial derivatives) of an output with respect to element values it is necessary to isolate δV_0 or δI_0 on the left-hand side of (6.24) by properly assigning source values in the adjoint network and simultaneously recognizing necessary constraints on the variations of the source variables in the original network. Source variables are eliminated by recognizing that for a current source $\delta I_S = 0$ while for a voltage source $\delta V_S = 0$. Concurrently, let $\phi_S = 0$ for a current source input and $\psi_S = 0$ for a voltage source output. Outputs are sampled across zero valued sources. That is, an output voltage is sampled across a zero valued current source where $\delta I_0 = 0$ and $\phi_0 = 1$ such that

$$\delta V_O = \sum_G V_G \, \psi_G \, \delta G + \sum_C j\omega V_C \, \psi_C \, \delta C + \sum_\Gamma \frac{1}{j\omega} V_\Gamma \, \psi_\Gamma \, \delta\Gamma + \sum_{G_T} V_I \, \psi_D \, \delta G_T \qquad (6.25)$$

where ϕ_0 is the output current source branch in the adjoint network. Similarly, an output current is sampled through a zero valued voltage source where $\delta V_0 = 0$ and $\psi_0 = -1$ such that

$$\delta I_O = \sum_G V_G \, \psi_G \, \delta G + \sum_C j\omega V_C \, \psi_C \, \delta C + \sum_\Gamma \frac{1}{j\omega} V_\Gamma \, \psi_\Gamma \, \delta\Gamma + \sum_{G_T} V_I \, \psi_D \, \delta G_T \qquad (6.26)$$

where ψ_0 is the output voltage source in the adjoint network. As can be inferred, for the adjoint network all of the original sources are set to zero while the network is excited by a source at the output. Individual element sensitivities are seen to be given by the product of the branch voltages in the original and adjoint networks where, except for the reversal of controlled sources, the original and adjoint networks are the same. The results obtained above are summarized in Table 6.1.

The results derived above can be extended to linear and nonlinear dc with the obvious provision that capacitors are replaced by open circuits and inductors by short-circuits. Note that for nonlinear circuits, the adjoint network is linear where nonlinear elements are replaced by linear small-signal conductances and transconductances as used in the Jacobian for Newton-Raphson iteration. In this case, all Norton equivalent current sources are treated as any other independent current source and set to zero. Sensitivity in the time domain can be derived in a manner similar to that used above[J8]; however, it can be shown that for the (linear) adjoint network, time runs backward. This added complication has limited the effectiveness of this approach. Time domain sensitivity is not considered further here.

Element	Original Branch Relation	Adjoint Branch Relation	Sensitivity
Conductance	$I_G = G\,V_G$	$\phi_G = G\,\psi_G$	$V_G\,\psi_G$
Capacitance	$I_C = j\omega C\,V_C$	$\phi_C = j\omega C\,\psi_C$	$j\omega V_C\,\psi_C$
Inductance	$I_\Gamma = \dfrac{\Gamma}{j\omega}\,V_\Gamma$	$\phi_\Gamma = \dfrac{\Gamma}{j\omega}\,\psi_\Gamma$	$\dfrac{1}{j\omega}\,V_\Gamma\psi_\Gamma$
Mutual Inductance	$I_1 = \dfrac{\Gamma_{11}}{j\omega}\,V_1 + \dfrac{\Gamma_M}{j\omega}\,V_2$ $I_2 = \dfrac{\Gamma_M}{j\omega}\,V_1 + \dfrac{\Gamma_{22}}{j\omega}\,V_2$	$\phi_1 = \dfrac{\Gamma_{11}}{j\omega}\,\psi_1 + \dfrac{\Gamma_M}{j\omega}\,\psi_2$ $\phi_2 = \dfrac{\Gamma_M}{j\omega}\,\psi_1 + \dfrac{\Gamma_{22}}{j\omega}\,\psi_2$	$\dfrac{V_1\psi_1}{j\omega} \quad \dfrac{V_2\psi_1}{j\omega}$ $\dfrac{V_1\psi_2}{j\omega} \quad \dfrac{V_2\psi_2}{j\omega}$
Voltage-Controlled Voltage Source	$V_D = A_V\,V_C$	$\phi_C = A_V\,\phi_D$	$V_C\,\phi_D$
Voltage-Controlled Current Source	$I_D = G_T\,V_C$	$\phi_C = G_T\,\psi_D$	$V_C\,\psi_D$
Current-Controlled Voltage Source	$V_D = R_T\,I_C$	$\psi_C = R_T\,\phi_D$	$I_C\,\phi_D$
Current-Controlled Current Source	$I_D = A_I\,I_C$	$\psi_C = A_I\,\psi_D$	$I_C\,\psi_D$

TABLE 6.1 - Summary of Adjoint Network Element Sensitivity

Computationally, the evaluation of the adjoint network response can proceed quite straightforwardly once all sources are properly taken into account. Clearly the same basic circuit equations as used in the original network can be used in the adjoint network. It has been observed, however, that certain network formulations allow even greater efficiency as will now be brought out for the case of nodal analysis.

In matrix form the nodal equations can be written

$$Y(j\omega)V = I \tag{6.27}$$

If LU decomposition is used in their solution, (6.27) can be rewritten as

$$L\,U\,V = I \tag{6.28}$$

where the dependence on $j\omega$ is still assumed but not explicitly shown. For the adjoint network element descriptions of Table 6.1, it can easily be shown that the admittance matrix of $\hat{\eta}$ is given by

$$\hat{Y} = Y^T \tag{6.29}$$

Thus, the adjoint network may be solved in stages by recognizing

$$Y^T\psi = \phi \tag{6.30}$$

$$(LU)^T\psi = \phi \tag{6.31}$$

$$U^T L^T\psi = \phi \tag{6.32}$$

such that

$$U^T\psi^* = \phi \tag{6.33}$$

$$L^T\psi = \psi^* \tag{6.34}$$

Since both L and U are triangular, $(L^T)^{-1}$ and $(U^T)^{-1}$ are easily computed resulting in ψ^* and then ψ. In fact ψ^* is computed from (6.33) by a single forward substitution step while ψ is computed from (6.34) by a single backward substitution step. This procedure is equivalent to transposing array indices in a program implementation. Where one used to proceed down a column, one now proceeds across a row and vice versa.

6.3 Small-Signal Sensitivities

For small-signal frequency response sensitivities, it is often the case that quantities like dB gains, magnitudes and phases are of interest. The sensitivity of these quantities can be computed as follows: For dB gain

$$G = 20 \log |V_O| \tag{6.35}$$

such that

$$\frac{\partial G}{\partial P} = \frac{20 \log e}{|V_O|} \frac{\partial |V_O|}{\partial P} \tag{6.36}$$

but

$$\frac{\partial |V_O|}{\partial P} = \frac{\partial}{\partial P} (V_R^2 + V_I^2)^{1/2} \tag{6.37}$$

$$= \frac{1}{2} (V_R^2 + V_I^2)^{-1/2} \frac{\partial}{\partial P} (V_R^2 + V_I^2) \tag{6.38}$$

$$= \frac{1}{2|V_O|} (2V_R \frac{\partial V_R}{\partial P} + 2V_I \frac{\partial V_I}{\partial P}) \tag{6.39}$$

$$= \frac{1}{|V_O|} \mathrm{Re}\,(\overline{V_O} \frac{\partial V_O}{\partial P}) \tag{6.40}$$

where V_R and V_I are the real and imaginary parts of V_O respectively, Re(...) means real part of, and $\overline{V_O}$ is the complex conjugate of V_O. Thus

$$\frac{\partial G}{\partial P} = \frac{20 \log e}{V_O^2} \mathrm{Re}\,(\overline{V_O} \frac{\partial V_O}{\partial P}) \tag{6.41}$$

$$= 20 \log e\ \mathrm{Re}\,(\frac{1}{V_O} \frac{\partial V_O}{\partial P}) \tag{6.42}$$

where P is any circuit element.

Similarly for phase,

$$\phi = K \tan^{-1} \frac{V_I}{V_R} \tag{6.43}$$

such that

$$\frac{\partial \phi}{\partial P} = K \left\{ 1 + \left[\frac{V_I}{V_R} \right]^2 \right\}^{-1} \frac{\partial (V_I/V_R)}{\partial P} \tag{6.44}$$

$$= K \frac{V_R^2}{V_O^2} \frac{V_R \dfrac{\partial V_I}{\partial P} - V_I \dfrac{\partial V_R}{\partial P}}{V_R^2} \tag{6.45}$$

$$= \frac{K}{V_O^2} \, \mathrm{Im} \left[\overline{V_O} \, \frac{\partial V_O}{\partial P} \right] \tag{6.46}$$

$$= \frac{57.29577}{V_O^2} \, \mathrm{Im} \left[\overline{V_O} \, \frac{\partial V_O}{\partial P} \right] \tag{6.47}$$

$$= 57.29577 \, \mathrm{Im} \left[\frac{1}{V_O} \, \frac{\partial V_O}{\partial P} \right] \tag{6.48}$$

These and other results are summarized in Table 6.2.

Where nonlinear circuits are operated in a small-signal mode at or about a dc operating point, the sensitivity of any output parameter with respect to a resistor value will be made up of several components in addition to a "direct" component. This fact results because a resistor variation can affect bias point which in turn can affect small-signal device parameters which influence small-signal performance. Under these conditions for a bipolar transistor,

$$\left. \frac{\partial V_O}{\partial P} \right|_{Total} = \frac{\partial V_O}{\partial P} + \sum_{Q} \left[\sum_{P_Q} \frac{\partial V_O}{\partial P_Q} \left(\frac{\partial P_Q}{\partial V_{BE}} \frac{\partial V_{BE}}{\partial P} + \frac{\partial P_Q}{\partial V_{BC}} \frac{\partial V_{BC}}{\partial P} \right) \right] \tag{6.49}$$

where the first term is the direct component, the first summation is over all nonlinear devices Q and the second summation is over all device parameters P_Q which depend on operating point. It can thus be seen that both dc and small-signal ac sensitivities are required.

$$\frac{\partial G}{\partial P} = 20 \log e \ \text{Re} \left(\frac{1}{V_O} \frac{\partial V_O}{\partial P} \right)$$

$$\frac{\partial \phi}{\partial P} = 57.29577 \ \text{Im} \left(\frac{1}{V_O} \frac{\partial V_O}{\partial P} \right)$$

$$\frac{\partial |V_O|}{\partial P} = \frac{1}{|V_O|} \ \text{Re} \left(\bar{V}_0 \frac{\partial V_O}{\partial P} \right)$$

$$\frac{\partial V_R}{\partial P} = \text{Re} \left(\frac{\partial V_O}{\partial P} \right)$$

$$\frac{\partial V_I}{\partial P} = \text{Im} \left(\frac{\partial V_O}{\partial P} \right)$$

TABLE 6.2 - Sensitivity of Small-Signal Gain and Phase Quantities

6.4 Noise and Group Delay Response

As previously mentioned, the adjoint network concept can be used to facilitate the calculation of noise and group delay responses. The noise response analysis proceeds from a consideration of only the output and source terms (6.6) which can be written as

$$\sum_S (\delta V_s \phi_s - \psi_s \delta I_s) + \delta V_O \phi_O - \psi_O \delta I_O = 0 \tag{6.50}$$

It is assumed that all noise sources are modeled as current sources with a quiescent (or nominal) value of zero. The excitation-response situation for the adjoint network is given by

$$\phi_s = 0 \qquad \phi_O = 1 \qquad \delta I_0 = 0 \tag{6.51}$$

such that (6.50) becomes

$$\delta V_O = \sum_S \psi_S \, \delta I_S \tag{6.52}$$

If δI_S is the mean square value of each noise source, then δV_0 from (6.52) becomes the mean square output noise voltage. In this application, only the branch voltages of each noisy element are required from the adjoint analysis.

Group delay is defined by

$$\tau = -\partial \phi(\omega)/\partial \omega \tag{6.53}$$

where $\phi(\omega)$ is the phase as a function of frequency ω. If the branch relations for the reactive elements of Table 6.1 are considered, it can be recognized that no computational difference exists between frequency as a parameter and an element value as a parameter. That is, in the adjoint network derivation considered previously, it is just as easy to attribute perturbations in the original network to frequency as to an element value. In this case (6.14) and (6.17) become

$$(I_C + \delta I_C) = j\,(\omega + \delta\omega)\,C\,(V_C + \delta V_C) \tag{6.54}$$

and

$$(I_\Gamma + \delta I_\Gamma) = \frac{\Gamma}{j\,(\omega + \delta\omega)}\,(V_\Gamma + \delta V_\Gamma) \tag{6.55}$$

such that

$$\delta I_C = j\,\delta\omega\,C\,V_C + j\,\omega\,C\,\delta V_C \tag{6.56}$$

and

$$\delta I_\Gamma = \frac{\Gamma}{j\,\omega}\,\delta V_\Gamma - \frac{\Gamma}{j\,\omega}\,V_\Gamma\,\frac{\delta\omega}{\omega} \tag{6.57}$$

The substitution of (6.56) and (6.57) into (6.6) accompanied with the element identification Table 6.1 and the proper excitation-response situation for the adjoint network results in

$$\delta V_O = \sum_C j\,C\,V_C\,\psi_C\,\delta\omega - \sum_\Gamma \frac{\Gamma}{j\,\omega}\,V_\Gamma\psi_\Gamma\,\frac{\delta\omega}{\omega} \tag{6.58}$$

Thus from (6.48) and (6.58), the group delay (6.53) is easily evaluated.

7. POLE-ZERO EVALUATION

7.1 Two-Port Transfer Functions

Linear pole-zero circuit analysis is considered as a separate topic because of the specialized techniques and problems involved relative to frequency domain analysis. Naturally, the poles and zeros of a transfer function, once computed, can provide the same frequency response information as a linear ac small-signal analysis performed by solving complex admittance equations in the frequency domain for $s = j\omega$. However, in many situations it is the poles and zeros themselves which are of primary interest to the circuit designer.

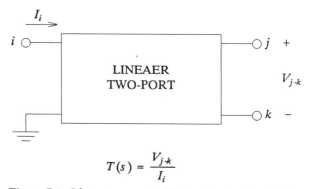

$$T(s) = \frac{V_{j \cdot k}}{I_i}$$

Figure 7.1 - Linear two-port with transfer function $T(s)$

A linear two-port network with input nodes i and *ground* and output nodes j and k is shown in Figure 7.1. To be as general as possible, it is assumed that the desired transfer function is

$$T(s) = (V_j - V_k)/I_i = V_{j \cdot k}/I_i \tag{7.1}$$

Assume that a nodal formulation is used such that the i th equation can be written as

$$\cdots + y_{ii} V_i + y_{ij} V_j + y_{ik} V_k + \cdots = I_i \tag{7.2}$$

A slight rearrangement of (7.2) results in

$$\cdots + y_{ii} V_i + y_{ij} (V_j - V_k) + (y_{ij} + y_{ik}) V_k + \cdots = I_i \tag{7.3}$$

In general, Cramer's rule can be used to express the transfer function $T(s)$ as the ratio of two determinants. Thus

$$T(s) = V_i/I_i = \frac{|Y_{ii}|}{|Y|} \quad j = i, \quad k = ground \tag{7.4}$$

where $|Y| = Det\ Y$ and $|Y_{ii}|$ is the minor obtained from $|Y|$ when the ith row and ith column are deleted. Similarly,

$$T(s) = V_j/I_i = \frac{|Y_{ij}|}{|y|} \quad j \neq i, \quad k = ground \tag{7.5}$$

Note that in (7.5), even if Y is symmetric in structure, Y_{ij} is very probably not. Furthermore, Y_{ij} may initially have a zero on its diagonal. For the case of $k \neq ground$, it can be seen from (7.3) that the jth column of Y must be added to the kth column and that the jth variable then represents $V_{j-k} = V_j - V_k$. In this case, $T(s)$ is given by

$$T(s) = V_{i-k}/I_i = \frac{|Y_{ii}^{k+i}|}{|Y|} \quad j = i,\ k \neq ground \tag{7.6}$$

and

$$T(s) = V_{j-k}/I_i = \frac{|Y_{ij}^{k+j}|}{|Y|} \quad j \neq i,\ k \neq ground \tag{7.7}$$

respectively where $|Y_{ii}^{k+i}|$ is the minor obtained from $|Y|$ with column i added to column k and then the ith row and column deleted with a similar manipulation for $|Y_{ij}^{k+j}|$.

For a voltage source input such that $I_i = 0$ and V_i represents the source, $|Y|$ is replaced by $|Y_{ii}|$ in the previous four equations. The above results are summarized in Table 7.1.

From Table 7.1 it is easily seen that for the zeros of $T(s)$, the possible determinants consist of

Case	Determinant
1	$|Y_{ii}|$
2	$|Y_{ij}|$
3	$|Y_{ii}^{k+i}|$
4	$|Y_{ij}^{k+j}|$

Current Input	$\dfrac{V_i}{I_i} = \dfrac{\mid Y_{ii} \mid}{\mid Y \mid}$	$\dfrac{V_j}{I_i} = \dfrac{\mid Y_{ij} \mid}{\mid Y \mid}$
	$\dfrac{V_{i\text{-}k}}{I_i} = \dfrac{\mid Y_{ii}^{k+i} \mid}{\mid Y \mid}$	$\dfrac{V_{j\text{-}k}}{I_i} = \dfrac{\mid Y_{ij}^{k+j} \mid}{\mid Y \mid}$
Voltage Input	$\dfrac{V_i}{V_i} = \dfrac{\mid Y_{ii} \mid}{\mid Y_{ii} \mid}$	$\dfrac{V_j}{V_i} = \dfrac{\mid Y_{ij} \mid}{\mid Y_{ii} \mid}$
	$\dfrac{V_{i\text{-}k}}{V_i} = \dfrac{\mid Y_{ii}^{k+i} \mid}{\mid Y_{ii} \mid}$	$\dfrac{V_{j\text{-}k}}{V_i} = \dfrac{\mid Y_{ij}^{k+j} \mid}{\mid Y_{ii} \mid}$

TABLE 7.1 - Two-Port Transfer Functions

With regard to these four cases, two generalizations can be made. The first is that cases 1 and 2 are special cases of 3 and 4 respectively where the negative output node is grounded (common with the negative input node). The second generalization is that cases 1 and 3 are special cases of 2 and 4 respectively where the positive output node is coincident with the positive input node. The conclusion to be derived from these two generalizations is that the most general case, Case 4, is all that needs to be handled.

The approach to be adopted then is the following:

1) Add the column of the positive output node (j) to the column of the negative output node (k).

2) Replace the column of the positive output node (j) with the column of the positive input node (i).

3) Delete the row and column of the positive input node (i).

The above sequence of operations will result in the correct determinant for the zeros of $T(s)$ for any of the four cases described previously.

To confirm that only the above cases need to be considered, the possible combinations of input and output node requests are detailed along with rearrangement rules in Table 7.2. It can be seen that the only unique cases are cases 1, 3 and the combination of 4 and 8. Each of these situations is handled by the previous general approach.

Case	Condition	Action = > Case
1	$i = j$, $k = ground$	
2	$i = j$, $k \neq ground$	$i \xrightarrow{\leftarrow} ground$, $j \xrightarrow{\leftarrow} k$ => 3
3	$i \neq j$, $k = ground$	
4	$i \neq j$, $k \neq ground$	
5	$i = k$, $k = ground$	$j \xrightarrow{\leftarrow} k$ => 1
6	$i = k$, $k \neq ground$	$i \xrightarrow{\leftarrow} ground$ => 3
7	$i \neq k$, $k = ground$	$j \xrightarrow{\leftarrow} k$ => 3
8	$i \neq k$, $k \neq ground$	

NOTE: Cases **4** and **8** = > floating output

TABLE 7.2 - Port Rearrangement Rules

It can be seen that once recognized, a multiplicity of possible port descriptions can be handled by a single, unified approach by following a few simple rules. As with floating voltage sources, etc. it is possible using the modified nodal formulation to simplify the above approach further. Since branch currents and node voltages are already handled with the modified nodal approach, the only remaining problem is branch voltages. More generally, the problem is the differences of node voltages. With this idea in mind, the equations can be augmented with another variable $V\pm$ such that the equation pattern looks like

	$V+$	$V-$	$V\pm$	RHS
$\varepsilon+$				
$\varepsilon-$				
$\varepsilon\pm$	-1	+1	+1	

where

$$V\pm \ = \ V+ \ - \ V- \tag{7.8}$$

7.2 Muller's Method

It was previously shown that the problem of determining the poles and zeros of a linear transfer function can be transformed into an equivalent problem -- that of finding the zeros of two determinants. A technique for finding such zeros which has proved particularly effective is the iterative root-finding method of Muller[11]. Originally developed for finding roots of polynomials, this method can be applied directly to $Det\ Y(s)$ and $Det\ Y_{ij}(s)$, etc. In Muller's method, the determinant is initially evaluated at three points and modeled by a quadratic equation. A zero of this quadratic is then used in place of one of the original evaluation points. This procedure is continued and a sequence of better and better approximations to each zero of the determinant is developed. As each new zero is found, its effect at each frequency is divided out of the determinant evaluation.

As formulated by Muller, this method proceeds in the following way: It is assumed that the zeros of

$$f(X) = a_n X^n \ + a_{n-1} X^{n-1} + \ \cdots \ + a_1 X \ + a_0 \tag{7.9}$$

are desired. The function $f(X)$ can be approximated by a quadratic

$$b_2 X^2 + b_1 X \ + b_0$$

where

$$\begin{bmatrix} X_i^2 & X_i & 1 \\ X_{i-1}^2 & X_{i-1} & 1 \\ X_{i-2}^2 & X_{i-2} & 1 \end{bmatrix} \begin{bmatrix} b_2 \\ b_1 \\ b_0 \end{bmatrix} = \begin{bmatrix} f(X_i) \\ f(X_{i-1}) \\ f(X_{i-2}) \end{bmatrix} = \begin{bmatrix} f_i \\ f_{i-1} \\ f_{i-2} \end{bmatrix} \tag{7.10}$$

Alternatively, define

$$h = X - X_i \qquad h_i = X_i - X_{i-1} \qquad h_{i-1} = X_{i-1} - X_{i-2} \tag{7.11}$$

and

$$\lambda = h/h_i = (X - X_i)/(X_i - X_{i-1})$$

$$\lambda_i = h_i/h_{i-1} = (X_i - X_{i-1})/(X_{i-1} - X_{i-2}) \tag{7.12}$$

and finally

$$\delta_i = 1 + \lambda_i = (X_i - X_{i-2})/(X_{i-1} - X_{i-2}) \tag{7.13}$$

A quadratic can then be written explicitly as

$$\left[f_{i-2}\lambda_i^2 - f_{i-1}\lambda_i\delta_i + f_i\lambda_i \right] \lambda^2$$

$$+ \left[f_{i-2}\lambda_i^2 - f_{i-1}\delta_i^2 - f_{i-1}\delta_i^2 + f_i(\lambda_i + \delta_i) \right] \lambda$$

$$+ \delta_i f_i$$

The zero of smallest magnitude of the above quadratic is chosen as λ from which

$$h = \lambda h_i \tag{7.14}$$

and

$$X_{i+1} = X_i + h \tag{7.15}$$

Muller's method is illustrated in Figure 7.2.

In solving the above quadratic for λ, Muller makes use of an alternative to the usual formula for the zeros of a quadratic. That is, given the quadratic equation

$$aX^2 + bX + c = 0 \tag{7.16}$$

then

$$a + b\frac{1}{X} + c\frac{1}{X^2} = 0 \tag{7.17}$$

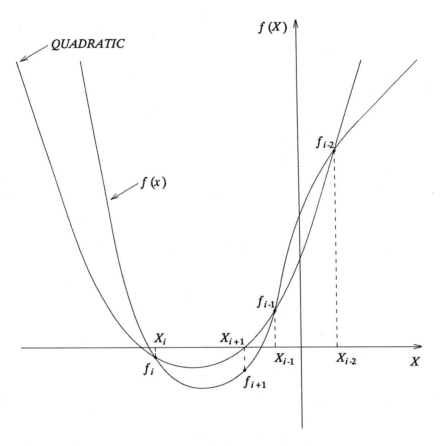

Figure 7.2 - Muller's method for real X where X_{i+1} is taken as smallest zero of quadratic approximation

such that

$$\frac{1}{X} = \frac{-b \pm \sqrt{b^2 - 4ac}}{2c} \tag{7.18}$$

or

$$X = \frac{-2c}{b \pm \sqrt{b^2 - 4ac}} \tag{7.19}$$

The latter formula (7.19) is used to solve for λ above where the smallest zero is chosen by maximizing the magnitude of the denominator.

For each zero, Muller suggests terminating the iteration sequence when

$$|h|/|X| < {\sim}10^{-7} \tag{7.20}$$

or when

$$|f(x)| < {\sim}10^{-20} \text{ or } 10^{-25} \tag{7.21}$$

Muller, for the case of polynomials, also suggests starting at

$$X_0 = -1, \quad X_1 = 1, \quad \text{and } X_2 = 0 \tag{7.22}$$

such that

$$h_1 = 2 \quad h_2 = -1 \quad \lambda_2 = \frac{-1}{2} \quad \delta_2 = \frac{1}{2} \tag{7.23}$$

This choice yields a starting quadratic in X

$$f(X) \approx a_2 X^2 + a_1 X + a_0 \tag{7.24}$$

which is a good approximation to $f(X)$ in the region near $X = 0$. Finally, note that as each new root is found, its effect is divided out of the original function $f(X)$ resulting in the reduced function

$$f^k(X) = \frac{f(X)}{\prod\limits_{j=1}^{k} (X - X_j)} \tag{7.25}$$

where k zeros $(X_1 \cdots X_j \cdots X_k)$ have been found. Successive zeros are found from $f^k(X)$ rather than $f(X)$.

In adapting Muller's method to the problem of finding the zeros of a determinant, several slight modifications must be adopted. First, consider the case of a linear active RC circuit where the general term in Y is of the form $G + sC$. The function $f(s)$ is then given by

$$f(s) = Det\ Y(s) = a_n s^n + a_{n-1} s^{n-1} + \cdots + a_0 \tag{7.26}$$

In principle, the expansion of $Det\ Y(S)$ into an nth order polynomial could be carried out such that Muller's method could be applied directly. It is not practical to do this however because of the number of accurate digits required for the polynomial coefficients. Instead, $Det\ Y(S)$ is evaluated directly from the nodal admittance matrix. The previously described procedure of diagonalizing Y via Gaussian elimination or LU factorization and then taking the product of the terms on the diagonal is used. This procedure is accurate and it takes advantage of and preserves the sparsity of the nodal admittance matrix. As before the effects of previous roots are always divided out as new roots are sought.

In the case of a polynomial, the order is known such that the total number of zeros, n, is also known. When Muller's method is applied directly to determinants, as described above, no *a priori* knowledge of the number of zeros exists. Consequently, a means of terminating the procedure after n zeros are found must be devised. As noted above, an expansion of $Det\ Y(s)$ into a polynomial results in (7.26). Similarly

$$\prod_{j=1}^{n} (s - s_j) = s^n + b_{n-1} s^{n-1} + \cdots + b_1 s + b_0 \tag{7.27}$$

Thus,

$$f^n(s) = \frac{a_n s^n + a_{n-1} s^{n-1} + \cdots + a_1 s + a_0}{s^n + b_{n-1} s^{n-1} + \cdots + b_1 s + b_0} = a_n \tag{7.28}$$

Equations (7.28) implies that when all zeros have been found, the reduced function $f^n(s)$ equals the constant a_n. This fact not only provides a stopping criteria but also results in a unique description of $Det\ Y(S)$ where

$$Det\ Y(s) = f^n(s) \prod_{j=1}^{n} (s - s_j) = a_n \prod_{j=1}^{n} (s - s_j) \tag{7.29}$$

Finally, the desired transfer function then can be written as

$$T(S) = \frac{a_{mz} \prod_{j=1}^{m} (s - s_j)}{a_{np} \prod_{j=1}^{n} (s - p_j)} \tag{7.30}$$

where it is assumed there are m zeros and n poles.

To this point, only linear active RC circuits have been considered. In the general case, linear active RLC circuits whose general term in $Y(s)$ is $sC + G + \dfrac{1}{sL}$ must be considered. Now

$$Det\ Y(s) = a_n s^n + \cdots + a_1 s + a_0 + \frac{a_{-1}}{s} + \cdots + \frac{a_{-m}}{s} \tag{7.31}$$

and $f^n(s)$ is no longer a constant. The problem is that neither n nor m are known *a priori* because of the possible redundancy due to loops of capacitors and cutsets of inductors. If m were known, it would be possible to start with the function

$$f(s) = s^m \ Det \ Y(s) = a_n \ s^{n+m} + \cdots + a_0 s^m + \cdots + a_{-m} \qquad (7.32)$$

in which case

$$f^{n+m}(s) = const = a_{-m} \qquad (7.33)$$

A procedure for determining m can be based on the following observation: For values of s near zero, (7.31) must blow-up while (7.32) remains continuous and well-defined. The following procedure is thus adopted. $Det \ Y(s)$ is evaluated at four points near zero as shown in Figure 7.3. From these points, the slope of $Det \ Y(s)$ on either side of the origin is evaluated. If there are no inductive roots, these slopes should be of the same sign, either both positive or both negative. In this case, $m = 0$ and $f(s) = Det \ Y(s)$. If the slopes are of opposite sign, there must be one or more inductive roots. In this case, the modified functions $s \ Det \ Y(s), \ldots s^m \ Det \ Y(s)$ are evaluated at the four points until a value of m is found which results in the same slopes on both sides of the origin. The function defined by (7.32) is then used in determining all remaining roots. After this procedure has been performed for both poles and zeros, the resulting transfer function can then be written as

$$T(s) = \frac{s^{np}}{s^{mz}} \frac{a_{-mz}}{a_{-np}} \frac{\prod(s - z_i)}{\prod(s - p_j)} \qquad (7.34)$$

$$= s^{np-mz} \frac{a_{-mz}}{a_{-np}} \frac{\prod(s - z_i)}{\prod(s - p_j)} \qquad (7.35)$$

Note that if $np > mz$, then $T(s)$ has zeros at the origin. Conversely, if $np < mz$ then $T(s)$ has poles at the origin (an unlikely occurrence). If $np = mz$, there are neither poles nor zeros at the origin due to inductive components.

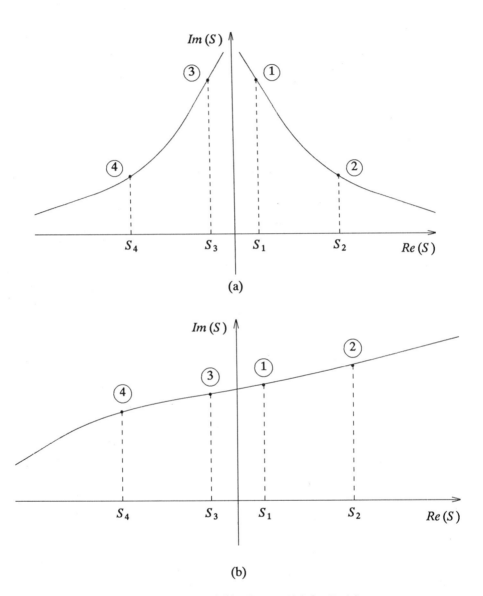

Figure 7.3 - (a) *Det* $Y(s)$ and (b) s^m *Det* $Y(s)$ for *Re* (s) near zero

8. STATISTICAL ANALYSIS

The intent statistical analysis or simulation is to obtain a statistical measure of circuit performance as a function of processing and fabrication variations. The Monte Carlo method of obtaining a statistical measure of performance involves repeated analyses where for each analysis element and parameter values are selected from their respective distributions. Typically a sample size of 100 or more simulations is required. This chapter will concentrate primarily on the development of a language and terminology for statistically describing circuits.

In order to provide a complete statistical description of electronic circuit elements used in integrated as well as discrete circuit designs, three fundamental capabilities must exist and be provided for within a program. The first is the ability to describe in natural terms a basic sample space or probability distribution function (PDF). The second is the ability to generate correlated random variables in such a way as to simulate parameter tracking. (e.g., the beta's of two integrated bipolar transistors from the same wafer would be expected to match more closely than those of devices from different wafers.) Finally, the third capability is to be able to combine random variables representing fundamental device parameters in such a way as to simulate correlation among derived or measurable parameters. The distinction being made here is between tracking of the same parameter in different devices and correlations among different parameters within the same device.

In what follows, each of these three capabilities is considered in more detail while a language and procedure for implementation are described as vehicles to illustrate the basic concepts.

8.1 Distributions

More important than the ability of a user to be able to describe a parameter or elements probability distribution function is the ability to be able to do so in the simplest and most natural of terms. The problem can be resolved into one of describing the shape of the distribution function (normal, uniform, etc.), its size (range, standard deviation, etc.), and the location of the nominal value within that distribution (mean, median, mode, etc.).

The first step, that of describing a distribution's shape might be accomplished through terminology of the form

$$DISTRIBUTION\ [dname\,] \begin{cases} UNIFORM \\ NORMAL \\ TABLE\ tname \\ FUNCTION\ fname \end{cases}$$

where the distribution name is optional and a choice is to be made between the various descriptions. Here, the distribution name, *dname*, if present is the name of a parameter or element with which the distribution is to be associated or, if none exists, simply the reference name assigned to the distribution. For example

$R1\ \ 10\ \ 0\ \ 1K$

$DISTRIBUTION\ (R1)\ =\ TABLE\ TDIF$

In certain cases where a name is implied by the context of the statement, the name may be omitted as for example when the two statements above are combined into

$R1\ \ 10\ \ 0\ \ 1K\ \ DISTRIBUTION\ \ TABLE\ \ TDIF$

The second step, that of describing a distribution's size, might be accomplished through any of the following statements:

$$RANGE \begin{cases} minval\ \ maxval \\ NOMINAL\ \ \text{-}percent\ \ \text{+}percent \end{cases}$$

$$\left.\begin{matrix} MEAN \\ MEDIAN \\ MODE \end{matrix}\right\} value \left\{\begin{matrix} SD1 \\ SD2 \\ SD3 \end{matrix}\right\} value$$

$TOLERANCE\ \ percent$

Note that in some cases, such as that of a table described in absolute units, steps one and two may be combined. In any case, the third step, that of specifying the nominal value within the defined distribution, might be carried out with a statement of the form

$$NOMINAL \begin{cases} MEAN \\ MEDIAN \\ MODE \\ value \end{cases}$$

Here, as a default, the value of the element or parameter associated with its initial description may be taken as nominal or, in the event that none exists, the median or mean may be used.

Several examples of the use of this terminology might be

DISTRIBUTION (BETA) = UNIFORM, RANGE = 50,200, NOMINAL = MEDIAN

DISTRIBUTION (VEARLY) = NORMAL, MEDIAN = 60, SD 1 = 2

DISTRIBUTION (CJE) = TABLE CJE, NOMINAL = MEDIAN

By way of further clarification, several common distribution functions are shown in Figure 8.1 where a number of terms used above are illustrated. The point to be made here is that no single set of descriptive parameters is best suited to handle all cases. Rather, any combination of the above descriptions must be allowed provided that it be self-consistent and complete. Where discrepancies arise, the most recently entered description should override previous descriptions with possibly a warning diagnostic being issued.

The statistical distributions thus described are intended to impose constraints on the range and frequency of sampled parameter values. For example, given a normal distribution, most sampled values should be close to the distributions mean. Most computer operating systems, however, provide only pseudo-random number generators which produce an approximately uniform distribution of values over some finite range, typically 0 to 1. Clearly, a means of transforming such a uniform distribution function into the required arbitrary distribution functions is needed. This transformation can be accomplished rather simply by first transforming the arbitrary distribution into a cumulative distribution function and then performing inverse interpolation. This process is shown pictorially in Figure 8.2. Note that the cumulative distribution function is nothing more than the integral of the required arbitrary distribution function normalized to unity.

8.2 Tracking and Pivots

Within the present context, tracking is specifically taken to mean the correlation or degree of matching between the same element and/or parameter within different circuits or devices. The key idea is that a global population or distribution of parameter values is composed of a number of local or sub-populations. During the course of a statistical or Monte Carlo analysis, the generation of correlated parameter values which track is analogous to the selection of a single value from the global population which then defines a nominal local sub-population from which all other values are selected.

Mathematically, this method of generating correlated random variables is known as the method of pivots, where the first value selected from the global distribution is the pivot. That is, correlated parameter values P_1 and P_2 are generated by the expressions

$$P_1 = P_0(1 + T_G(1-\lambda)X_1 + \lambda X_0)) \qquad (8.1)$$

$$P_2 = P_0(1 + T_G(1-\lambda)X_2 + \lambda X_0)) \qquad (8.2)$$

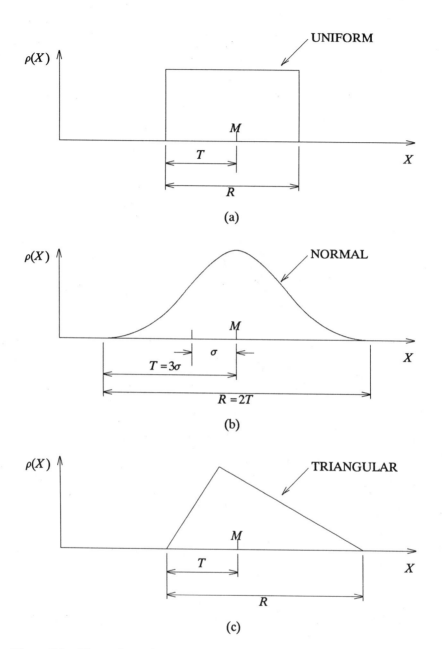

Figure 8.1 - Illustration of median (M), tolerance (T), range (R), and standard deviation (σ) for various distribution functions

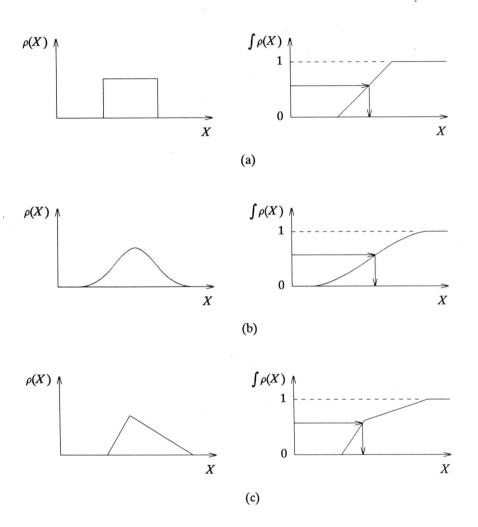

Figure 8.2 - Mapping from uniform to arbitrary distributions via cumulative distribution function

where it is assumed that $-1 \leq X_i \leq 1$ for a distribution whose range is symmetric. Here, P_0 is the assumed nominal or average parameter value with an associated global tolerance T_G, λ is a correlation coefficient or tracking factor as brought out below, X_0 is the pivot selected from the global distribution, while X_1 and X_2 are selected from the local or sub-distribution where it has been assumed that all distributions have been normalized.

The tracking factor λ can be given a physical interpretation as follows: Assume that associated with the local or sub-population is a tolerance T_L and define λ as

$$\lambda = 1 - T_L/T_G \tag{8.3}$$

Since it must be true that $T_L/T_G \leq 1$, $0 \leq \lambda \leq 1$. Further, the condition $T_L = T_G$ implies no tracking in which case $\lambda = 0$ while the condition $T_L = 0$ (the local population consists of a single value) implies perfect tracking in which case $\lambda = 1$. With this interpretation, the expressions for P_1 and P_2 above may be re-written as

$$P_1 = P_0(1 + T_L X_1 + (T_G - T_L)X_0) \tag{8.4}$$

$$P_2 = P_0(1 + T_L X_2 + (T_G - T_L)X_0) \tag{8.5}$$

Several special cases are of interest. For $X_0 = X_1 = X_2 = 0$,

$$P_1 = P_2 = P_0 \tag{8.6}$$

the nominal value, while for $X_0 = X_1 = X_2 = \pm 1$,

$$P_1 = P_2 = P_0(1 \pm T_G) \tag{8.7}$$

where it should be observed that the parameter values are constrained to lie within the range of the global distribution. For the case of perfect tracking ($\lambda = 1$, $T_L = 0$)

$$P_1 = P_2 = P_0(1 + T_G X_0) \tag{8.8}$$

while for the case of no tracking whatsoever ($\lambda = 0$, $T_L = T_G$)

$$P_1 = P_0(1 + T_L X_1) \tag{8.9}$$

$$P_2 = P_0(1 + T_L X_2) \tag{8.10}$$

as desired.

An interpretation of the approach just described in terms of integrated circuits might be made by considering the local populations to represent the variations of a parameter, such as β for a bipolar transistor, within a single chip while the global population represents the variation across the entire wafer and consequently includes many chips. However a random sample of integrated circuits usually includes chips from many different wafers as well. Thus, an even larger population exists, and a three step selection process is required (i.e. pick a wafer, pick a chip, pick a device). In this case a nested model which makes use of a pivot and a sub-pivot is required. That is,

$$P = P_0 \left\{ 1 + T_W \left[(1 - \lambda_{CW}) \left[(1 - \lambda_{DC}) X_D + \lambda_{DC} X_C \right] + \lambda_{CW} X_W \right] \right\} \qquad (8.11)$$

$$= \left\{ T_D (X_D - X_C) + \left[T_C (X_C - X_W) + (1 + T_W X_W) \right] \right\} \qquad (8.12)$$

$$= P_0 \left[1 + T_D X_D + (T_C - T_D) X_C + (T_W - T_C) X_W \right] \qquad (8.13)$$

Here, T_D, T_C, and T_W represent the tolerance on all devices within a chip, all devices within a wafer, and all devices respectively. The tracking factors are given by

$$\lambda_{DC} = 1 - T_D / T_C \qquad (8.14)$$

$$\lambda_{CW} = 1 - T_C / T_W \qquad (8.15)$$

where $T_D \leq T_C \leq T_W$. Finally, X_D, X_C, and X_W are random numbers chosen from normalized probability distribution functions representing the variation of parameter values within a chip, within a wafer and within all wafers respectively.

Again, it is useful to consider several special cases. For perfect tracking between devices within a chip, $(\lambda_{DC} = 1, T_D = 0)$

$$P = P_0 (1 + T_C X_C + (T_W - T_C) X_W) \qquad (8.16)$$

while for perfect tracking between chips as well $(\lambda_{DC} = \lambda_{CW} = 1, T_D = T_C = 0)$

$$P = P_0 (1 + T_W X_W) \qquad (8.17)$$

Finally, for no tracking whatsoever $(\lambda_{DC} = \lambda_{CW} = 0, T_D = T_C = T_W)$

$$P = P_0 (1 + T_D X_D) \qquad (8.18)$$

$$\equiv P_0 (1 + T_W X_D) \qquad (8.19)$$

Again, all parameter values lie within the range of the global distribution. Other combinations are handled as easily.

In the previous section, a terminology was proposed for describing a single distribution such as for representing parameter values of a device within a chip, of many devices within a wafer, and of all devices within all wafers. From the third form of the general tracking expression above (8.13), it is apparent that once each distribution is normalized, the only additional information required is the associated tolerances of each distribution and a pointer to the proper pivot and/or sub-pivot. The tolerances, of course, are available once each distribution is completely defined.

A way in which pivots may be established is to point to associated distributions with a pivot declaration of the form

PIVOT dname

Thus, for example, for the case of a single pivot and no sub-pivot, the following statements might be used:

DIST GLOBAL NORMAL MEDIAN 100 *SD* 3 15

DIST LOCAL NORMAL MEDIAN 100 *SD* 3 5 *PIVOT GLOBAL*

or for the case of device, chip and wafer tracking (pivot and sub-pivot)

DIST WAFER UNIFORM RANGE 50 200 *NOMINAL MEDIAN*

DIST CHIP TABLE TCHIP PIVOT WAFER

DIST DEVICE NORMAL MEDIAN 150 *SD* 1 10 *PIVOT CHIP*

The question may arise as to how to combine the individual statistics for the variation of a parameter across each chip into some kind of average or composite distribution representing a typical chip or even a wafer. One simple-minded approach might be to normalize the distribution associated with each individual parameter with respect to some measure of average such as mean, median or mode and then to combine these normalized distributions into an overall composite distribution which represents a typical chip. For variations across a wafer, the process could be repeated, this time without regard to chip. Finally, the entire population would be taken without regard to wafer or chip. For more detailed work, composites might be obtained by more sophisticated statistical methods.

8.3 Pivot Sorting Algorithm

A procedure for storing and sorting a collection of distributions might proceed as follows:

1) Search all distributions for those without pivot references and store those in order of occurrence as set 1.

2) Search remaining distributions for all those referencing distributions in set 1 as pivots and store in order of occurrence as set 2.

3) Finally, search remaining distributions for all those referencing set 2 and store in order of occurrence as set 3.

4) Any remaining pivots are in error. Note that during sorting, checks on tolerance magnitudes should be made to insure proper nesting.

By way of example, suppose the following are read in:

DIST PD 11 *NORMAL MEDIAN* 100 *SD* 1 5 *PIVOT PC* 1

DIST PD 12 *NORMAL MEDIAN* 100 *SD* 1 8 *PIVOT PC* 1

DIST PC 1 *NORMAL MEDIAN* 100 *SD* 1 10 *PIVOT PW*

DIST PD 21 *NORMAL MEDIAN* 100 *SD* 1 8 *PIVOT PC* 2

DIST PC 2 *NORMAL MEDIAN* 100 *SD* 1 10 *PIVOT PW*

DIST PW NORMAL MEDIAN 100 *SD* 1 15

Then the following sort would result

Name	Tol	Ptr	Xval	Pval
PW	TW	0	X_W	$V_W = 1 + T_W X_W$
PC 1	T_{C1}	PW	X_{C1}	$V_{C1} = V_W + T_{C1}(X_{C1} - X_W)$
PC 2	T_{C1}	PW	X_{C2}	$V_{C2} = V_W + T_{C2}(X_{C2} - X_W)$
PD 11	T_{C2}	PW	X_{C2}	$V_{C2} = V_W + T_{C2}(X_{C2} - X_W)$
PD 12	T_{D12}	PC 1	X_{D12}	$V_{D11} = V_{C1} + T_{D11}(X_{D11} - X_{C1})$
PD 21	T_{D21}	PC 2	X_{D21}	$V_{D21} = V_{C2} + T_{D21}(X_{D21} - X_{C2})$

where it should be noted from (8.12) that by computing and storing the intermediate quantities

$$V_W = 1 + T_W X_W \tag{8.20}$$

$$V_C = V_W + T_C (X_C - X_W) \tag{8.21}$$

$$V_D = V_C + T_D (X_D - X_W) \tag{8.22}$$

that P can be calculated in stages as

$$P = P_0 \left\{ T_D \left(X_D - X_C \right) + \left[T_C \left(X_C - X_W \right) + \left(1 + T_W X_W \right) \right] \right\}$$

$$= P_0 \left\{ T_D \left(X_D - X_C \right) + \left[T_C \left(X_C - X_W \right) + \left(\quad V_W \quad \right) \right] \right\}$$

$$= P_0 \left\{ T_D \left(X_D - X_C \right) + \left(\quad V_C \quad \right) \right\}$$

$$= P_0 \left\{ \quad V_D \quad \right\}$$

Note also that values must be computed for each element or parameter which points to a particular pivot. Thus if n transistors point to a pivot for β, n random numbers must be evaluated from that pivot resulting in n values or random factors. Consequently, for each pivot pointed to by a parameter or element, a multiplicity factor must be determined.

8.4 Correlation

As pointed out previously, correlation within the present context is taken to mean the inter-dependence of different parameters within the same device or circuit element. In general, this type of dependence is not nearly so easy to incorporate as simple parameter tracking since complex mathematical relationships may be involved. By way of example consider the following: From simple, first-order theory transistor betas and reverse saturation currents are known to vary inversely with basewidth while forward and reverse minority-carrier transit times are known to vary as basewidth squared. Since basewidth is subject to statistical variations from device to device, a method of conveying the type of dependence just described, or any other dependence, to the simulation program is required.

Assume the basewidth (W) is a parameter whose statistical distribution is known and given by

$$W = W_N \left[1 + T_D X_D + \left(T_C - T_D \right) X_C + \left(T_W - T_C \right) X_W \right] \tag{8.23}$$

$$= W_N \left(1 + \delta_W \right) \tag{8.24}$$

where δ_W represents the fractional variation of basewidth from its nominal value. If the dependence of β on basewidth cited above is assumed, β may be written as

$$\beta = \beta_N (W_N / W) \tag{8.25}$$

$$= \beta_N \left[W_N / \left(W_N (1 + \delta_W) \right) \right] \tag{8.26}$$

$$= \beta_N / (1 + \delta_W) \tag{8.27}$$

where β_N is the nominal value of β. Thus, the statistical dependence of β on basewidth can be obtained by a multiplication of its nominal value by the reciprocal of one plus the fractional deviation in basewidth δ_W. The latter term is readily described and computed via the models and terminology of the previous several sections. What is needed to complete the statistical description is β is a means of specifying this desired dependence. A possible terminology might be based on the use of the word "*VARIATION*" to refer to quantities such as $1 + \delta_W$. Thus

$$W = W_N * VAR(W) \tag{8.28}$$

and

$$\beta = \beta_N / VAR(W) \tag{8.29}$$

where

$$VAR(W) = 1 + \delta_W \tag{8.30}$$

and *VAR* is an abbreviation for *VARIATION*.

The variational quantity as defined above can be used in conjunction with other operators such as

$$+ \quad * \quad / \quad EXP \quad ALOG \quad etc.$$

to describe parameter correlations as illustrated by the following example:

$$DIST \quad W = \cdots$$

$$DIST \quad BF = BF \ / \ VAR(W)$$

$$DIST \quad IS = IS \ / \ VAR(W)$$

$$DIST \quad TF = TF \ * \ VAR(W) \ * \ VAR(W)$$

As another example, consider the built-in potential, ϕ, associated with a pn-junction and given by

$$\phi = V_T \ ln \ (N_d N_a / n_i^2) \tag{8.31}$$

where V_T is the thermal voltage, N_d and N_a are the donor and acceptor ion concentrations at the junction, and n_i is the intrinsic carrier concentration. If statistical descriptions are supplied for both N_d and N_a, then ϕ can be written as

$$\phi = \phi_N + VT \ln \left[VAR(Nd) * VAR(Na) \right] \qquad (8.32)$$

where ϕ_N is the nominal value of ϕ. The statistical description supplied would be

$DIST\ (ND) = \ldots$

$DIST\ (NA) = \ldots$

$DIST\ (PHI) = PHI + VT * ALOG \left[VAR(ND) * VAR(NA) \right]$

Of the topics presented in this chapter, the problem of determining an accurate correlation model among the various parameters within a semiconductor device model is perhaps the most difficult. For bipolar devices, work has been done by Divekar, *et. al.* [K9] in developing empirical models based upon a statistical technique known as Factor Analysis. In addition, the ASTAP [C11] has for many years been used for statistical simulations.

REFERENCES

A. GENERAL CAD

A1. F.F. Kuo and J.F. Kaiser, *System Analysis by Digital Computer,* Wiley, 1965.

A2. G.T. Herskowitz, *Computer Aided Integrated Circuit Design,* McGraw-Hill, 1968.

A3. L.P. Huelsman, *Digital Computations in Basic Circuit Theory,* McGraw-Hill, 1968.

A4. F.F. Kuo and W.G. Magnuson, Jr., *Computer Oriented Circuit Design,* Prentice-Hall, 1969.

A5. B.J. Ley, *Computer Aided Analysis and Design for Electrical Engineers,* Holt, Rinehart, & Winston, 1970.

A6. D.A. Calahan, *Computer-Aided Network Design, Rev. Ed.,* McGraw-Hill, 1972.

A7. L.O. Chua and P.M. Lin, *Computer-Aided Analysis of Electronic Circuits: Algorithms and Computational Techniques,* Prentice-Hall, 1975.

B. NUMERICAL ANALYSIS AND PROGRAMMING

B1. R.W. Hamming, *Numerical Methods for Scientists and Engineers,* McGraw-Hill, 1962.

B2. A.S. Householder, *The Theory of Matrices in Numerical Analysis,* Blaisdell, 1964.

B3. L. Fox, *An Introduction to Numerical Linear Algebra,* Oxford Univ. Press, 1965, 327 pp.

B5. E. Isaacson and H.B. Keller, *Analysis of Numerical Methods,* Wiley, 1966, 541 pp.

B6. C.E. Froberg, *Introduction to Numerical Analysis,* Addison-Wesley, 2nd. Ed., 1969, 433 pp.

B7. B.W. Arden and K.N. Astill, *Numerical Algorithms: Origins and Applications,* Addison-Wesley, 1970, 308 pp.

B8. C. Daniel and F.S. Wood, *Fitting Equations to Data,* Wiley, 1971, 342 pp.

B9.   A. Ralston, *Introduction to Programming and Computer Science,* McGraw-Hill, 1971, 513 pp.

B10. M.L. Stein and W.D. Munro, *Introduction to Machine Arithmetic,* Addison-Wesley, 1971, 295 pp.

B11. B.W. Kernighan and P.J. Plauger, *The Elements of Programming Style,* McGraw-Hill, 1974, 174 pp.

B12. E. Yourdan, *Techniques of Program Structure and Design,* Prentice-Hall, 1975, 364 pp.

C. CIRCUIT SIMULATION

C1.   L.D. Milliman, et. al., "Circus, A Digital Computer Program for Transient Analysis of Electronic Circuits," *Rep. 346-2,* Diamond Ordnance Fuze Lab, Washington, D.C., January 1967.

C2.   R.W. Jensen and M.D. Lieberman, *IBM Electronics Circuit Analysis Program: Techniques and Applications,* Prentice-Hall, 1968.

C3.   W.J. McCalla and W.G. Howard Jr., "BIAS-3--A Program for the Nonlinear DC Analysis of Bipolar Transistor Circuits," *IEEE J. Solid-State Circuits,* SC-6, Feb. 1971, pp. 14-19.

C4.   I.A. Cermak and D.B. Kirby, "Nonlinear Circuits and Statistical Design," *Bell System Tech J.,* Vol. 50, April 1971, pp. 1173-1197.

C5.   F.H. Branin, G.R. Hogsett, B.L. Lunde and L.E. Kugel, "ECAP-II--A New Electronic Circuit Analysis Program," *IEEE J. Solid-State Circuits,* Vol. SC-6, August 1971, pp. 146-166.

C6.   T.E. Idleman, F.S. Jenkins, W.J. McCalla and D.O. Pederson, "SLIC--A Simulator for Linear Integrated Circuits," *IEEE J. Solid-State Circuits,* Vol. SC-6, August 1971, pp. 188-203.

C7.   F.S. Jenkins and S.P. Fan, "TIME--A Nonlinear DC and Time-Domain Circuit Simulation Program," *IEEE J. Solid-State Circuits,* SC-6, August 1971, pp. 182-188.

C8.   L.W. Nagel and R.A. Rohrer, "Computer Analysis of Nonlinear Circuits Excluding Radiation (CANCER)," *IEEE J. Solid-State Circuits,* SC-6, August 1971, pp. 166-182.

C9.   L.W. Nagel and D.O. Pederson, "SPICE--Simulation Program with Integrated Circuit Emphasis," *Memo No. ERL-M382,* Electronics Research Laboratory, University of California, Berkeley, April, 1973.

C10. V.K. Manaktala, "A Versatile Small Circuit Analysis Program," *IEEE Trans. Circuit Theory,* CT-20, Sept. 1973, pp. 583-586.

C11. W.T. Weeks, A.J. Jimenez, G.W. Mahoney, D. Mehta, H. Qassemzadeh and T.R. Scott, "Algorithms for ASTAP-- A Network Analysis Program," *IEEE Trans. Circuit Theory,* CT-20, November 1973, pp. 628-634.

C12. T.K. Young and R.W. Dutton, "An MOS Simulation for Integrated Nonlinear Circuits with Modular Built-In Model," *Tech. Rpt. No. 5010-1,* Stanford Electronics Laboratories, Stanford University, Stanford, California, July 1974.

C13. L.W. Nagel, "SPICE2: A Computer Program to Simulate Semiconductor Circuits," *Memo No. ERL-M520,* Electronics Research Laboratory, University of California, Berkeley, May, 1975.

C14. H. Kop, P. Chuang, A. Lachner and B. McCalla, "SLIC--A Comprehensive Nonlinear Circuit Simulation Program," Conference Record, Ninth Annual Asilomar Conference on Circuits, Systems and Computers, November 1975, pp. 348-353.

C15. T.K. Young and R.W. Dutton, "MINI--MSINC--A Minicomputer Siomulator for MOS Circuits with Modular Built-in Model," *Tech. Rpt. No. 5013-1,* Stanford Electronics Laboraties, Stanford University, Stanford, California, March 1976.

C16. R.W. Jensen and L.P. McNamee, *Handbook of Circuit Analysis Languages and Techniques,* Prentice-Hall, 1976, pp. 809.

D. ELEMENT AND MACRO-MODELS

D1. C.A. Desoer and E.S. Kuh, *Basic Circuit Theory,* McGraw-Hill, 1969.

D2. M.A. Murray-Lasso, "Black-Box Models for Linear Integrated Circuits," *IEEE Trans. Education,* E-12, September 1969, pp. 170-180.

D3. R.B. Yarbrough, "Circuit Models for Transformers," *IEEE Trans. Education,* E-12, September 1969, pp. 181-188.

D4. E.B. Kozemchak, "Computer Analysis of Digital Circuits by Macromodeling," *IEEE International Symposium on Circuit Theory,* Los Angeles, California, April 1972.

D5. N.B. Rabbat, W.D. Ryan and S.Q. Hossian, "Macro-modelling and Transient Simulation in Integrated Digital Systems," *IEE Proc. Int'l Conf. on Computer Aided Design,* University of Southhampton, April 1972, pp. 263-270.

D6. E.F. Bernard, "A Transmission-Line Model for Transient CACD Programs," *IEEE J. Solid-State Circuits,* Vol. SC-7, June 1972, pp. 270-273.

D7. P. Balaban, "Calculation of the Capacitance Coefficients of Planar Conductors on a Dielectric Surface," *IEEE Trans. Circuit Theory,* CT-20, November 1973, pp. 725-737.

E. EQUATION FORMULATION

E1.   G. Kron, "A Method of Solving Very Large Physical Systems in Easy Stages," *Proc. IRE,* Vol. 42, April 1954, pp. 680-686.

E2.   T.R. Bashkow, "The A Matrix, New Network Description," *Trans. IRE Circuit Theory,* CT-4, September 1957, pp. 117-119.

E3.   P.R. Bryant, "The Order of Complexity of Electrical Networks," *Proc. Inst. Elec. Eng.,* Monograph 335E, Vol. 106C, June 1959, pp. 174-188.

E4.   P.R. Bryant, "The Explicit Form of Bashkow's A-Matrix," *Trans. IRE Circuit Theory,* CT-9, September 1962, pp. 303-306.

E5.   R.L. Wilson and W.A. Masena, "An Extension of Bryant- Bashkow A-Matrix," *IEEE Trans. Circuit Theory,* CT-12, March 1965, pp. 120-122.

E6.   E.S. Kuh and R.A. Rohrer, "The State-Variable Approach to Network Analysis," *Proc. IEEE,* 53, July 1965, pp. 672-686.

E7.   J.A.C. Bingham, "A Method of Avoiding Loss of Accuracy in Nodal Analysis," *Proc. IEEE,* Vol. 55, March 1967, pp. 409-410.

E8.   F.H. Branin, Jr., "Computer Methods of Network Analysis," *Proc. IEEE,* 55, November 1967, pp. 1787-1809.

E9.   R.E. Parkin, "A State Variable Method of Circuit Analysis Based on a Nodal Approach," *Bell System Tech J.,* November 1968, pp. 1957-1970.

E10.   G.D. Hachtel, R.K. Brayton and F.G. Gustavson, "The Sparse Tableau Approach to Network Analysis and Design," *IEEE Trans. Circuit Theory,* CT-18, January 1971, pp. 101-118.

E11.   P.M. Russo and R.A. Rohrer, "The Three-Link Approach to the Time Domain Analysis of a Class of Nonlinear Networks," *IEEE Trans. Circuit Theory,* CT-18, May 1971, pp. 400-403.

E.12. C.W. Ho, A.E. Ruehli and P.A. Brennan, "The Modified Nodal Approach to Network Analysis," *IEEE Trans. Circuits and Systems,* CAS-22, June 1975, pp. 504-509.

E13.   J.P. Freret, "Minicomputer Calculation of the DC Operating Point of Bipolar Circuits," *Tech. Rept. No. 5015-1,* Stanford Electronics Laboratories, Stanford University, Stanford, California, May 1976.

E14.   J.P. Freret and R.W. Dutton, "Successful Circuit Simulation Using Minicomputers," 19th Midwest Symposium for Circuits and Systems, Milwaukee, Wisconsin, August 1976.

F. NONLINEAR DC ANALYSIS

F1.   D.F. Davidenko, "On a New Method of Numerical Solution of Systems of Nonlinear Equations," *Pokl. Akad.*, Nauk SSSR, 88, 1953, pp. 601-602.

F2.   C.G. Broyden, "A Class of Methods for Solving Nonlinear Simultaneous Equations," *Math. Comput.*, 19, October 1965, pp. 577-593.

F3.   F.H. Branin, Jr. and H.H. Wang, "A Fast Reliable Iteration Method for dc Analysis of Nonlinear Networks," *Proc. IEEE*, 55, November 1967, pp. 1819-1826.

F4.   C.G. Broyden, "A New Method of Solving Nonlinear Simultaneous Equations," *Comp. J.*, 12, February 1969, pp. 94-99.

F5.   G.C. Brown, "DC Analysis of Nonlinear Networks," *Electron Lett.*, 5, August 1969, pp. 374-375.

F6.   H. Shichman, "Computation of DC Solutions for Bipolar Transistor Networks," *IEEE Trans. on Circuit Theory,"* CT-16, November 1969, pp. 460-466.

F7.   R.A. Rohrer, "Successive Secants in the Solution of Nonlinear Equations," *AMS,* 1970, pp. 103-112.

F8.   J.M. Ortega and W.R. Rheinholdt, *Iterative Solution of Non-linear Equations in Several Variables,* Academic Press, New York, 1970.

F9.   I.A. Cermak, "DC Solution of Nonlinear State Space Equations in Circuit Analysis," *IEEE Trans. Circuit Theory,"* CT-18, March 1971, pp. 312-314.

F10. I.A. Cermak and D.B. Kirby, "Nonlinear Circuits and Statistical Design, *Bell System Tech J.,* 50, April 1971, pp. 1173-1195.

F11. W.H. Kao, "Comparison of Quasi-Newton Methods for the DC Analysis of Electronic Circuits", *Coordinated Sci. Lab. Rep. R-577,* University of Illinois, Urbana-Champaign, July, 1972.

G. SPARSE MATRICES TECHNIQUES

G1.   H.M. Markowitz, "The Elimination Form of the Inverse and Its Application to Linear Programming," *Management Sci.,* 3, April 1957, pp. 255-269.

G2.   N. Sato and W.F. Tinney, "Techniques for Exploiting the Sparsity of the Network Admittance Matrix," *IEEE Trans.* (Power App. Syst), 82, December 1963, pp. 944-950.

G3.   W.F. Tinney and J.W. Walker,"Direct Solutions of Sparse Network Equations by Optimally Ordered Triangular Factorization," *Proc. IEEE,* 55, November 1967, pp. 1801-1809.

G4. J.F. Pinel and M.L. Blostein,"Computer Techniques for the Frequency Analysis of Linear Electrical Networks," *Proc. IEEE,* 55, November 1967, pp. 1810-1819.

G5. A. Schneider, "Estimate of the Number of Arithmetic Operations Required in LU-Decomposition of a Sparse Matrix," *IEEE Trans. Circuit Theory,* CT-17, May 1970, pp. 269-270.

G6. R.D. Berry, "An Optimal Ordering of Electronic Circuit Equations for a Sparse Matrix Solution," I IEEE Trans. Circuit Theory, CT-18, January 1971, pp. 139-146.

G7. R.S. Norin and C. Pottle, "Effective Ordering of Sparse Matrices Arising from Nonlinear Electrical Networks," *IEEE Trans Circuit Theory,* CT-18, January 1971, pp. 139-146.

G8. H.Y. Hsieh and M.S. Ghausi, "On Optimal-Pivoting Algorithms in Sparse Matrices," *IEEE Trans. Circuit Theory,* CT-19, January 1972, pp. 93-96.

G9. A.M. Erisman and G.E. Spies, "Exploiting Problem Characteristics in the Sparse Matrix Approach to Frequency Domain Analysis," *IEEE Trans. Circuit Theory,* CT-19, May 1972, pp. 260-264.

G10. H.Y. Hsieh and M.S. Ghausi, "A Probabilistic Approach to Optimal Pivoting and Prediction of Fill-in for Random Sparse Matrices," *IEEE Trans. Circuit Theory,* CT-19, July 1972, pp. 329-336.

G11. D.J. Rose and R.A. Willoughby, *Sparse Matrices and Their Applications,* Plenum Press, 1972, 215 pp.

G12. B. Danbert and A.M. Erisman, "Hybrid Sparse Matrix Methods," *IEEE Trans. Circuit Theory,* CT-20, November 1973, pp. 641-649.

G13. H.Y. Hsieh, "Pivoting-Order Computation Method for Large Random Sparse Systems," *IEEE Trans. Circuits and Systems,* CAS-21, March 1974, pp. 225-230.

G14. H.Y. Hsieh, "Fill-in Comparisons Between Gauss-Jordan and Gaussian Elimination," *IEEE Trans. Circuits and Systems,* CAS-21, March 1974, pp. 230-233.

G15. M. Hakhla, K. Singhal and J. Vlach, "An Optimal Pivoting Order for the Solution of Sparse Systems of Equations,' *IEEE Trans. Circuits and Systems,* CAS-21, March 1974, pp. 230-233.

H. NUMERICAL INTEGRATION

H1. A. Nordsieck, "On Numerical Integration of Ordinary Differential Equations," *Math. Comp,* 16, 1962, pp. 22-49.

H2. C.G. Dahlquist,"A Special Stability Problem for Linear Multistep Methods," *BIT.* 3, 1963, pp. 27-43.

H3. H.H. Rosenbrock, "Some General Implicit Processes for the Numerical Solution of Differential Equations," *Comput. J.,* 5, January 1963, pp. 329-330.

H4. D.A. Pope, "An Exponential Method of Numerical Integration of Ordinary Differential Equations," *Comm ACM,* 6, August 1963, pp. 491-493.

H5. C.W. Gear, "NUmerical Integration of Stiff Ordinary Equations," University of Illinois at Urbana-Champaign, Report 221, January 1967.

H6. M.E. Fowler and R.M. Warten, "A Numerical Technique for Ordinary Differential Equations with Widely Separated Eigenvalues," *IBM J. Res. Development,* September 1967, pp. 537-543.

H7. I.W. Sandberg and H. Shichman, "Numerical Integration of Systems of Stiff Nonlinear Differential Equations," *Bell Sys Tech J.,* 47, April 1968.

H8. C.W. Gear, "The Control of Parameters in the Automatic Integration of Ordinary Differential Equations," University of Illinois at Urbana-Champaign, File No. 757, May 1968.

H9. R.H. Allen, "Numerically Stable Explicit Integration Techniques Using a Linearized Range Kutta Extension", Report No. 39, Information Sciences Laboratory, Boeing Scientific Research Laboratory, October 1969.

H10. W. Liniger and R.A. Willoughby, "Efficient Integration Methods for Stiff Systems of Ordinary Differential Equations," *SIAM J. Numerical Anal.,* 7, March 1970, pp. 47-66.

H11. H. Shichman, "Integration System of a Nonlinear Network-Analysis Program," *IEEE Trans. Circuit Theory,* CT-17, August 1970, pp. 387-386.

H12. D.A. Calahan, "Numerical Considerations for Implementation of a Nonlinear Transient Circuit Analysis Program," *IEEE Trans. Circuit Theory,* CT-18, January 1971, pp. 66-73.

H13. C.W. Gear, "Simultaneous Numerical Solution of Differential-Algebraic Equations," *IEEE Trans. Circuit Theory,* CT-18, January 1971, pp. 89-95.

H14. C.W. Gear, *Numerical Initial Values Problems in Ordinary Differential Equations,* Prentice-Hall, 1971, 253 pp.

H15. R.K. Brayton, F.G. Gustavson and G.D. Hachtel, "A New Efficient Algorithm for Solving Differential- Algebraic Systems Using Implicit Backward Differentiation Formulas," *Proc. IEEE,* 60, January 1972, pp. 98-108.

H16. Y. Genin, "A New Approach to the Synthesis of Stiffly Stable Linear Multistep Formulas," *IEEE Trans. Circuit Theory,* CT-20, July 1973, pp. 352-360.

H17. Z. Fazarinc, "Designer Oriented CAD," *IEEE Trans. Circuit Theory,* CT-20, November 1973, pp. 673-682.

H18. M.L. Blostein, "The Transient Analysis of Nonlinear Electronic Networks," *IEEE Short Course Notes*, 735C06, 1973.

H19. W.M. van Bokhoven, "Linear Implicit Differentiation Formulas of Variable Step and Order," *IEEE Trans. Circuits and Systems*, CAS-22, February 1975, pp. 109-115.

H20. A.H. Stroud, *Numerical Quadrature and the Solution of Ordinary Differential Equations*, Springer-Verlag.

I. POLES AND ZEROS

I1. D.E. Muller, "A Method for Solving Algebraic Equations Using an Automated Computer," in *Mathematical Tables and Other Aids to Computation (MTAC)*, 10, 1956, pp. 208-215.

I2. J.G.F. Francis, "The Q-R Transformation-I," *Comput. J.*, 4, October 1961, pp. 265-271 and "The Q-R Transformation-II," *Comput. J.*, 4, January 1962, pp. 332-345.

I3. R.W. Brockett, "Poles, Zeros and Feedback; State-Space Interpretation," *IEEE Trans. Automatic Control*, AC-10, April 1965, pp. 129-135.

I4. J.H. Wilkinson, *The Algebraic Eigenvalue Problem*, Oxford, 1965.

I5. B.N. Parlett, "The LU and QR Algorithms," in *Mathematical Methods for Digital Computers*, Vol. 2, (A. Ralston and H.S. Wilf, Eds.), Wiley, 1967, pp. 116-130.

I6. I.W. Sandberg and H.C. So, "A Two-Sets- of-Eigenvalues Approach to the Computer Analysis of Linear Systems," *IEEE Trans. Circuit Theory*, CT-16, November 1969, pp. 509-517.

I7. C. Pottle, "A 'Textbook' Computerized State-Space Network Analysis Program," *IEEE Trans. Circuit Theory*, CT-16, November 1969, pp. 566-568.

I8. D.A. Calahan and W.J. McCalla, *In Sparse Matrices and Their Applications*, (D.J. Rose and R.A. Willoughby, Eds.), Plenum Press, 1972.

I9. Y.M. Wong and C. Pottle, "On the Sparse Matrix Computation of Critical Frequencies," *IEEE Trans. Circuits and Systems*, CAS-23, February 1976, pp. 91-95.

J. SENSITIVITY AND OPTIMIZATION

J1. R. Fletcher and M.J.D. Powell, "A Rapidly Convergent Descent Method for Minimization," *Comput. J.*, 6, June 1963, pp. 163-168.

J2. J.A. Nelder and R. Mead, "A Simplex Method for Function Minimization," *Comp. J.,* 7, 1965, pp. 308.

J3. S. Narayanan, "Transistor Distortion Analysis Using Volterra Series Representation," *Bell System Tech J.,* May-June 1967.

J4. G.C. Temes and D.A. Calahan, "Computer-Aided Network Optimization--The State-of-the-Art," *Proc. IEEE,* 55, November 1967, pp. 1832-1863.

J5. R.A. Rohrer, "Fully-Automated Network Design by Digital Computer: Preliminary Considerations," *Proc. IEEE,* November 1967, pp. 1929-1939.

J6. S.W. Director and R.A. Rohrer, "Interreciprocity and its Implications," *Proc. International Symposium Network Theory,* Belgrad, September 1968, pp. 11-30.

J7. J.S. Kowalik and M.R. Osborne, *Methods for Unconstrained Optimization Problems,* American Elsevier Publishing Co.,New York, 1968, pp. 148.

J8. S.W. Director and R.,A. Riohrer, "The Generalized Adjoint Network and Network Sensitivities," *IEEE Trans. Circuit Theory,* CT-16, August 1969, pp. 318-323.

J9. S.W. Director and R.A. Rohrer, "Automated Network Design--The Frequency Domain Case," *IEEE Trans. Circuit Theory,* CT-16, August 1969, pp. 330-337.

J10. S.W. Director and R.A. Rohrer, "On the Design of Resistance n-Port Networks by Digital Computer," *IEEE Trans. Circuit Theory,* CT-16, August 1969, pp. 337-346.

J11. S.W. Director, "Survey of Circuit-Oriented Optimization Techniques," *IEEE Trans. Circuit Theory,* CT-18, January 1971, pp. 3-10.

J12. A.J. Broderson, S.W. Director and W.A. Bristol, "Simultaneous Automated AC and DC Design of Linear Integrated Circuit Amplifiers," *IEEE Trans. Circuit Theory,"* CT-18, January 1971, pp. 50-58.

J13. R.I. Dowell and R.A. Rohrer, "Automated Design of Biasing Circuits," *IEEE Trans. Circuit Theory,* CT-18, January 1971, pp. 85-89.

J14. C.W. Ho, "Time-Domain Sensitivity Computation of Networks Containing Transmission Lines," *IEEE Trans. Circuit Theory,* CT-18, January 1971, pp. 114-122.

J15. B.A. Wooley, "Automated Design of DE-Coupled Monolithic Broad-Band Amplifiers," *IEEE J. Solid-State Circuits,* SC-6, February 1971, pp. 24-34.

J16. R.A. Rohrer, L.W. Nagel, R.G. Meyer and L. Weber, "Computationally Efficient Electronic Circuit Noise Calculations," *IEEE J. Solid-State Circuits,* SC-6. August 1971, pp. 204-213.

J17. T.J. Apprille and T.N. Trick, "Steady State Analysis of Nonlinear Circuits and Periodic Inputs," *Proc. IEEE,* 60, January 1972, pp. 108-114.

J18. R.G. Meyer, M.J. Shensa and R. Eschenback, "Cross Modulation and Intermodulation in Amplifiers at High Frequencies", *IEEE J. Solid-State Circuits*, SC-7, February 1972, pp. 16-23.

J19. T.J. Aprille Jr., and T.N. Trick, "A Computer Algorithm to Determine the Steady-State Response of Nonlinear Oscillations," *IEEE Trans. Circuit Theory*, CT-19, July 1972, pp. 354-360.

J20. D.A. Wayne, S.W. Director and A.J. Broderson, "Automated Design of Large-Signal Amplifiers for Minimum Distortion," *IEEE Trans. Circuit Theory*, CT-19, September 1972, pp. 531-533.

J21. S.L.S. Jacoby, J.S. Kowalik and J.T. Pizzo, *Iterative Methods for Nonlinear Optimization Problems*, Prentice-Hall, 1972, pp. 274.

J22. D. Agnew, "Iterative Improvement of Network Solutions," *Proc. Sixteenth Midwest Symposium on Circuit Theory*, Waterloo, Ontario, April 1973.

J23. R.G. Meyer, L.W. Nagel and S.K. Lui, "Computer Simulation of 1/f Noise Performance of Electronic Circuits," *IEEE J. Solid-State Circuits*, SC-8, June 1973, pp. 237-240.

J24. S. Narayanan and H.C. Poon, "An Analysis of Distortion in Bipolar Transistors Using Integral Charge Control Model and Volterra Series," *IEEE Trans. Circuit Theory*, CT-20, July 1973, pp. 341-351.

J25. F.R. Colon and T.N. Trick, "Fast Periodic Steady-State Analysis for Large Signal Electronic Circuits," *IEEE J. Solid-State Circuits*, SC-8, August 1973, pp. 260-269.

J26. S.W. Director, W.A. Bristol and A.J. Broderson "Fabrication-Based Optimization of Linear Integrated Circuits," *IEEE Trans. Circuit Theory*, CT-20, November 1973, pp. 690-697.

J27. Y.L. Kuo, "Distortion Analysis of Bipolar Transistor Circuits," *IEEE Trans. Circuit Theory*, CT-20 November 1973, pp. 709-717.

J28. S.H. Chisholm and L.W. Nagel, "Efficient Computer Simulation of Distortion in Electronic Circuits," *IEEE Trans. Circuit Theory*, CT-20, November 1973, pp. 742-745.

J29. W.A. Schuppan, "A Comparison of the Gaussian Elimination Method and an Optimization Technique Using the Adjoint Network for the Analysis of Linear Circuits," University of Illinois at Urbana-Champaign, *MS Thesis*, 1973.

J30. R.P. Brent, *Algorithms for Minimization Without Derivatives*, Prentice-Hall, 1973, pp. 195.

J31. O. Wing and J.V. Behar, "Circuit Design by Minimization Using the Hessian Matrix," *IEEE Trans. Circuits and Systems*, CAS-21, September 1974, pp. 643-649.

J32. A.J. Broderson and S.W. Director, "Computer Evaluation of Differential Amplifier Performance," *IEEE Trans. Circuits and Systems,* CAS-21, November 1974, pp. 735-742.

J33. T.N. Trick, F.R. Colon and S.P. Fan, "Computation of Capacitor Voltage and Inductor Current Sensitivities with Respect to Initial Conditions for the Steady-State Analysis of Nonlinear Periodic Circuits," *IEEE Trans. Circuits and Systems,* CAS-22, May 1975, pp. 391-396.

J34. G.D. Hachtel, M.R. Lightner and H.J. Kelly, "Application of the Optimization Program AOP to the DEsign of Memory Circuits," *IEEE Trans. Circuit Theory,* CT-22, June 1975, pp. 496-503.

J35. R.K. Brayton and S.W. Director, "Computation of Delay Time Sensitivities for Use in Time Domain Optimization," *IEEE Trans. Circuits and Systems,* CAS-22, December 1975, pp. 910-920.

K. STATISTICAL AND TOLERANCE ANALYSIS

K1. E.M. Butler, "Realistic Design Using Large-Change Sensitivities and Performance Contours," *IEEE Trans. Circuit Theory,* CT-18, January 1971, pp. 58-66.

K2. Special Issue on Statistical Circuit Design, *Bell System Tech J.,* April 1971, pp. 1099-1407.

K3. V.K. Manaktala and G.L. Kelly, "Computer-Aided Worst Case Sensitivity Analysis of Electrical Networks over a Frequency Interval," *IEEE Trans. Circuit Theory,* CT-19, January 1972, pp. 91-93.

K4. J.F. Pinel and K.A. Roberts, "Tolerance Assignment in Linear Networks Using Nonlinear Programming," *IEEE Trans. Circuit Theory,* CT-19, September 1972, pp. 475-480.

K5. A.R. Thorbjornsen and S.W. Director, "Computer- Aided Tolerance Assignment for Linear Circuits with Correlated Elements," *IEEE Trans. Circuit Theory,* CT-20, September 1973, pp. 518-524.

K6. J.W. Bandler and P.C. Lin, "Automated Network Design with Optimal Tolerances," *IEEE Trans. Circuit and Systems,* CAS-21, March 1974, pp. 219-222.

K7. P. Balaban and J.J. Golembeski, "Statistical Analysis For Practical Circuit Design," *IEEE Trans. Circuits and Systems,* CAS-22, February 1975, pp. 100-109.

K8. J.W. Bandler, P.C. Lin and H. Troup, "A Nonlinear Programming Approach to Optimal Design Centering, Tolerancing and Turning," *IEEE Trans. Circuits and Systems,* CAS-23, March 1976, pp. 155-165.

K9. D. Divekar, R.W. Dutton, and W.J. McCalla, "Experimental Study of Gummel-Poon Model Parameter Correlations for Bipolar Junction Transistors," *IEEE J. Solid State Circuits,* SC-12, October 1977, pp. 552-559.

INDEX

A

absolute stability 108
adjoint network(s) 125-135
A-stability 108-112
auxiliary function limiting 60-65

B

back substitution 17,19,21
backward difference formulas 92,97
backward Euler integration 88,91-92,
  94,99,101-102,104,108-111
BDF's 92,97,101
Berry 47-50
bi-directional threaded list 39-45
Bingham 31-32
branch constitutive equations (BCE)
  1,2,12
Brayton 100
Brown 60
Broyden 60

C

capacitor 2,87-90,92
Cermak 60,66
characteristic polynomial 109
chord method 70-74
Colon 64-65
consistent, integration method 99
construction, integration formulas
  96-101
convergence, Newton-Raphson iteration
  58-66
correlated random variables 151
correlation 158-160

Crout's method 25
cumulative distribution 151,153
current-controlled current source 2
current-controlled voltage source 2
current source 2

D

Dahlquist 108,111
Davidenko 60
determinants 29
difference equation(s) 88,108-110
distributions 149-151
Divekar 160
divided difference 101,104-105
Doolittle's method 23-25

E

eigenvalue 108-112
explicit integration method(s) 90

F

factor analysis 160
fill-ins 45-51,74
fixed bound limiting 60-65
floating voltage sources 8-9
forward Euler integration 87,91,94,
  101-102,104,108-111
forward substitution 21
Freret 29-30
functional iteration 53